MICROSCOPY HANDBOOKS 38

Confocal Laser Scanning Microscopy

Royal Microscopical Society MICROSCOPY HANDBOOKS

01 An Introduction to the Optical Microscope (2nd Edn)
03 Specimen Preparation for Transmission Electron Microscopy of Materials
04 Histochemical Protein Staining Methods
08 Maintaining and Monitoring the Transmission Electron Microscope
09 Qualitative Polarized Light Microscopy
11 An Introduction to Immunocytochemistry (Revised Edn)
15 Dictionary of Light Microscopy
16 Light Element Analysis in the Transmission Electron Microscope
17 Colloidal Gold: a New Perspective for Cytochemical Marking
18 Autoradiography: a Comprehensive Overview
19 Introduction to Crystallography (2nd Edn)
20 The Operation of Transmission and Scanning Electron Microscopes
21 Cryopreparation of Thin Biological Specimens for Electron Microscopy
22 An Introduction to Surface Analysis by Electron Spectroscopy
23 Basic Measurement Techniques for Light Microscopy
24 The Preparation of Thin Sections of Rocks, Minerals and Ceramics
25 The Role of Microscopy in Semiconductor Failure Analysis
26 Enzyme Histochemistry
27 *In Situ* Hybridization: a Practical Guide
28 Biological Microtechnique
29 Flow Cytometry
30 Food Microscopy: a Manual of Practical Methods, Using Optical Microscopy
31 Scientific PhotoMACROgraphy
32 Microscopy of Textile Fibres
33 Modern PhotoMICROgraphy
34 Contrast Techniques in Light Microscopy
35 Negative Staining and Cryoelectron Microscopy: the Thin Film Techniques
36 Lectin Histochemistry
37 Introduction to Immunocytochemistry (2nd Edn)
38 Confocal Laser Scanning Microscopy

Confocal Laser Scanning Microscopy

C.J.R. Sheppard
Department of Physical Optics, School of Physics,
and Australian Key Centre for Microscopy and Microanalysis,
University of Sydney, NSW 2006, Australia
Tel: +61 2 9351 5971/2553; fax: + 61 2 9351 7727;
e-mail: colin@Physics.usyd.edu.au

D.M. Shotton
Cell Biology Research Group,
Department of Zoology,
University of Oxford,
Oxford OX1 3PS, UK
Tel: +44 1865 271193/271234; fax: +44 1865 310447;
e-mail: david.shotton@zoo.ox.ac.uk

In association with the Royal Microscopical Society

C.J.R. Sheppard and D.M. Shotton
respectively Department of Physical Optics, School of Physics, and Australian Key Centre for Microscopy and Microanalysis, University of Sydney, NSW, Australia; and Cell Biology Research Group, Department of Zoology, University of Oxford, Oxford, UK

Published in the United States of America, its dependent territories and Canada by arrangement with BIOS Scientific Publishers Ltd, 9 Newtec Place, Magdalen Road, Oxford OX4 1RE, UK

© BIOS Scientific Publishers Limited, 1997

First published 1997

All rights reserved. No part of this book may be reproduced or transmitted, in any form or by any means, without permission.
The use of general descriptive names, trade names, trademarks, etc., in this publication, even if the former are not especially identified, is not to be taken as a sign that such names, as understood by the Trade Marks and Merchandise Marks Act, may accordingly be used freely by anyone.
While the advice and information in this book are believed to be true and accurate at the date of going to press, neither the authors nor the editors nor the publisher can accept any legal responsibility for any errors or omissions that may be made. The publisher makes no warranty, express or implied, with respect to the material contained herein.

A CIP catalogue record for this book is available from the British Library.

Library of Congress Cataloging-in-Publication Data
Sheppard, Colin.
    Confocal laser scanning microscopy / Colin Sheppard, David Shotton.
      p.   cm.
    Includes bibliographical references (p.) and index.
    ISBN 0-387-91514-1 (hardcover : alk. paper)
    1. Confocal microscopy. I. Shotton, David. II. Title.
QH224.S49 1998
502'.8'2--dc21   97-24717
                                                                                                      CIP

ISBN 0 387 91514 1 Springer-Verlag New York Berlin Heidelberg SPIN 19900576

Springer-Verlag New York Inc.
175 Fifth Avenue, New York,
NY 10010-7858, USA

Production Editor: Priscilla Goldby.
Typeset by Poole Typesetting (Wessex) Ltd, Bournemouth, UK.
Printed by Information Press Ltd, Eynsham, Oxon, UK.

Front cover: Confocal multiparameter image of a *Drosophila* embryo at the blastoderm stage, fluorescently triple-labelled for the transcription products of the genes *Hairy*, *Krüpple* and *Giant*. Details and credits as for *Figure 5.6* (Colour Plate IV).

Contents

Abbreviations	ix
Preface	xi

1. Introduction   1

Advantages of confocal scanning optical microscopy	1
The scanned image	3
Principle of out-of-focus blur rejection in the confocal fluorescence microscope	5
Methods of scanning	6
Single-beam scanning	6
Video-rate single-beam SOMs	7
Slit scanning	10
Multiple-beam scanning	10
Design of single-beam confocal laser scanning microscopes	10

2. Image formation in the confocal laser scanning microscope   15

Principles of image formation in non-confocal and confocal single-point scanning microscopes	15
Confocal axial scanning, height profiling and vertical sectioning	18
Other imaging modes, including non-confocal and confocal transmission microscopy, confocal reflection DIC, darkfield, polarization and interference microscopy, and optical-beam-induced current (OBIC) imaging	22
Photodetectors and imaging noise	24
Theoretical considerations of photodetector noise	24
Confocal photodetectors in practice	26
Choice and design of objectives, and avoidance of aberrations	27
Setting up and aligning a confocal microscope	30

3.	**Performance of confocal microscopes**	33

Confocal optical sectioning and enhancement of axial
  resolution   33
Use of the theoretical PSF to calculate the axial image
  intensity of a point object   33
Spatial frequency cutoff   34
Integrated intensity, and axial images of points and planes   34
Variation of axial resolution with numerical aperture   36
Axial resolution and spherical aberration   38
Lateral resolution enhancement   39
Reduced image degradation due to light scattering and
  autofluorescence   42
Sensitivity, geometric linearity, dynamic range and
  photometric accuracy   44

4.	**Image processing of confocal image data sets**	45

Computational requirements   45
Use of pseudocolour in multiparameter imaging   47
Three-dimensional image processing   47
Stereoscopic imaging   48
Surface and volume rendering   50
Computer animation of 3D and 4D confocal images   50

Colour plates   52

5.	**Confocal fluorescence microscopy**	61

Instrumentation for multiparameter confocal
  fluorescence imaging   61
Fluorescent preparation techniques for biological specimens   62
Blur-free optical sectioning of single cells: applications
  in cell and molecular biology   64
Applications in development biology, neurobiology and
  diagnostic cytopathology   66
Multiple wavelength imaging   67
Fluorescence ratio imaging of physiological parameters   67
Combined confocal epi-fluorescence and non-confocal
  transmission imaging   68
Time-resolved fluorescence   69
Conclusion   70

6.	**Biological applications of confocal reflection microscopy**	**71**
	Colloidal gold imaging	71
	Confocal reflection imaging of surface replicas and gold-coated SEM specimens	72
	Interference reflection contrast imaging	73
7.	**Industrial applications of confocal microscopy**	**75**
	Micrometrology	75
	Surface profiling and surface examination	76
	Thin film profiling	78
	Microscopy of semiconductor materials and devices	79
8.	**The future of confocal microscopy**	**83**
	Super-resolution	83
	Near-field scanning optical microscopy	84
	4 pi and theta confocal microscopy	85
	Two-photon excitation and other non-linear confocal techniques	87
	Conclusion	91
	Appendices	**93**
	Appendix A: Further reading	93
	Appendix B: Suppliers	97
	Appendix C: World Wide Web sites relevant to confocal microscopy	99
	Index	**101**

Abbreviations

2D	two-dimensional
3D	three-dimensional
4D	four-dimensional
CCD	charge coupled device
CLSM	confocal laser scanning microscope/microscopy
CMOS	complementary metal-oxide-semiconductor
CSOM	confocal scanning optical microscope/microscopy
DABCO	1, 4 diazobicyclo 2, 2, 2 octane
DIC	differential interference contrast
DPX	a proprietary hardening mountant for light microscope specimens (from BDH Merck, Product Number 36029). An alternative product is Mowiol (from Aldrich Chemical Co., Product Number 32,459–0)
FWHM	full width of a curve at the half-maximum height
Gbyte	gigabyte
GFP	green fluorescent protein from the jellyfish *Aequorea victoria*
MDCK	Madin–Darby canine kidney epithelial cells
Mbyte	megabyte
MOS	metal-oxide-semiconductor
NA	numerical aperture ($n \sin \alpha$)
NEA	negative electron affinity
NSOM	near-field scanning optical microscope/microscopy
OBIC	optical-beam-induced current
OTF	optical transfer function
PC	personal computer (either IBM-compatible or Apple Macintosh)
PIN	positively doped/intrinsic/negatively doped
PMT	photomultiplier tube
PSF	point spread function
RAM	random access memory
SEM	scanning electron microscope/microscopy
SNR	signal-to-noise ratio
SOM	scanning optical microscope/microscopy
STM	scanning tunnelling microscope/microscopy
TEM	transmission electron microscope/microscopy
UV	ultraviolet
VALAP	a 1:1:1 mixture of vaseline, lanoline and paraffin wax (melting point 51–53°C), which melts at about 65°C and is suitable for sealing coverslips bearing living cells to microscope slides

Preface

Confocal microscopy is a new science. While the idea for a confocal microscope was first patented by Minsky in 1957, and the first purely analogue mechanical confocal microscope was designed and produced by Eggar and Petran a decade later, it was not until the late seventies, with the advent of affordable computers and lasers, and the development of digital image processing software, that the first single-beam confocal laser scanning microscopes were developed in a number of laboratories and applied to biological and materials specimens. For biologists, a crucial turning point came in 1985 with the publication of six papers from four separate laboratories independently demonstrating the power of the confocal fluorescence microscope to eliminate out-of-focus blur, and thus to obtain three-dimensional (3D) data from intact biological specimens by non-invasive optical sectioning. In a remarkably rapid development, in which both of us were involved, the first commercial confocal laser scanning fluorescence microscopy systems were produced within 2 years of these publications by a small Oxfordshire company which has now become Bio-Rad Microscience Ltd. A number of other companies, including all the major microscope manufacturers worldwide, were swift to produce their own confocal instruments of widely varying designs and capabilities. During the last decade the availability of confocal laser scanning microscopes of ever-increasing power and sophistication has revolutionized the science of microscopy as applied to cell and developmental biology, physiology, cytogenetics, diagnostic pathology, and the material sciences. In particular, the ability to obtain a time-series of three-dimensional images from a living specimen, with temporal and spatial resolutions as good as or superior to video microscopy, has opened up new avenues of investigations previously impossible to contemplate.

Over the last decade, several excellent books and reviews on confocal microscopy have been published, but there has been a noticeable gap in the availability of a small handbook that introduces the interested student or research worker to this important microscopical technique, and that illustrates how it might benefit their own research. It is this gap that we hope this Handbook, the latest in the Royal Microscopical Society Microscopy Handbooks series, will fill.

Starting from first principles, this Handbook explains to the reader what a confocal microscope is, how it is constructed and used, what its benefits are, and why its imaging performance is superior to that of a conventional optical microscope. It discusses multiparameter confocal fluorescence microscopy, describes digital image processing and animation of 3D confocal images, illustrates applications of confocal microscopy in both the biomedical and the materials sciences, and concludes with a discussion of future developments in this new area of microscopy. Richly illustrated with colour micrographs and diagrams chosen for their clarity and didactic quality, this book also contains an up-to-date bibliography of the most informative publications on confocal microscopy, a catalogue of World Wide Web sites of relevance, and a listing of the names and addresses of confocal microscope and fluorescence filter manufacturers, image processing software vendors, and reagent suppliers. We hope that you will find it useful.

Colin Sheppard
David Shotton

Acknowledgements

The authors are most grateful to Bio-Rad Microscience Ltd for sponsoring the inclusion of colour plates in this RMS Handbook, and to Anna Smallcombe of Bio-Rad Microscience Ltd and numerous other colleagues named individually in the figure legends who have generously supplied and given permission for their micrographs to be used to illustrate the principles and practice of confocal microscopy.

1 Introduction

1.1 Advantages of confocal scanning optical microscopy

Confocal scanning optical microscopy (CSOM) has recently emerged as a significant new technique which exhibits several advantages over conventional optical microscopy. The most important of these stem from the fact that out-of-focus blur is essentially absent from confocal images, giving the capability for direct non-invasive serial optical sectioning of intact and even living specimens. This leads to the possibility of generating three-dimensional (3D) images of thick transparent objects such as biological cells and tissues. In a similar way, it allows profiling of the surfaces of 3D objects and multi-layer structures such as integrated circuits deposited on silicon, again by a non-contacting and non-destructive method.

In conventional epi-illumination light microscopy, the simultaneous illumination of the entire field of view of a specimen (full-field illumination) (*Figure 1.1a*) will excite fluorescence emissions or reflections throughout the whole depth of the specimen, rather than just at the focal plane. Much of the light collected by the objective lens to form the image will thus come from regions of the specimen above and below the selected focal plane, contributing as out-of-focus blur to the final image, and seriously degrading it by reducing contrast and sharpness. This explains why studies by conventional immunofluorescence microscopy of biological specimens are most informative on well-flattened cultured cells, and why the maximum usable thickness of frozen tissue sections for immunofluorescence observations is about 10 µm. Although less severe than in fluorescence microscopy, similar image degradation from out-of-focus regions of the specimen is also experienced in conventional transmission imaging modes, particularly phase-contrast microscopy.

However, by the very simple expedients of restricting the illumination of the specimen to a single point (or an array of points), which is scanned to produce a complete image (point scanning illumination) (*Figure 1.1b*), and of inserting a confocal imaging aperture in the optical system, as explained in Section 1.3, almost all the light emanating from regions above and below the focal plane of a confocal light microscope is physically prevented from contributing to the observed image, which thus (to a first approximation) contains only in-focus information. This endows the confocal microscope with significant axial resolution, and permits its use for serial non-invasive optical sectioning, and the acquisition of 3D image data.

Confocal microscopy has other additional advantages over conventional optical microscopy, including a small but significant improvement in lateral resolution, detailed in Chapter 3. It rejects stray light not only from the out-of-focus specimen planes but also light scattered from within the optical instrument itself, resulting in increased contrast and signal-to-noise ratio in the final image. Confocal microscopy is also compatible with computer image storage techniques, allowing, for example, generation of high-resolution digitized data sets of the 3D distribution of labels within cells or tissue, or of the topography of a surface, suitable for subsequent image processing (discussed in Chapter 4).

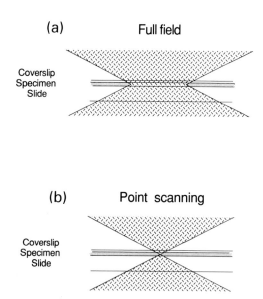

Figure 1.1. A comparison of the illumination experienced by the specimen during (a) full-field illumination in a conventional fluorescence microscope and (b) single-point illumination at the focal plane by a diffraction-limited point of light from an illuminating aperture in a single-beam laser scanning optical microscope, both employing an oil-immersion objective lens of high numerical aperture. The volume occupied by the specimen between the coverslip and the microscope slide is stippled, and the cone of illuminating light is cross-hatched. For simplicity, refraction effects are not shown. Reproduced by permission of the Royal Microscopical Society from Shotton, D.M. (1988) The current renaissance in light microscopy. II. Blur-free optical sectioning of biological specimens by confocal scanning fluorescence microscopy. *Proc. R. Microscop. Soc.* **23**: 289–297.

1.2 The scanned image

All forms of electronic light microscopy (i.e. light microscopy in which the optical information is converted into electronic form for subsequent processing and display) involve raster scanning of the electronic signal readout. However, such scanned image microscopy may be divided into two fundamentally distinct types. Firstly, *image-plane scanning* encompasses all forms of microscopy in which the two-dimensional (2D) optical image of the specimen is focused conventionally on to the image plane of an electronic imaging device, such as the faceplate of a video camera image tube, or the 2D semiconductor imaging array of a charge coupled device (CCD) camera, in which captured photons generate electron-hole pairs. After the simultaneous exposure of all areas of the electronic imaging device to the incoming optical image, the information is read out in serial electronic form, by the raster scanning of the faceplate or the sequential charge transfer of the CCD array, as a series of sequential image lines. Secondly, and in contrast, scanning optical microscopy (SOM) involves *object-plane scanning*. In this, a 2D optical image of the specimen is *not* formed within the microscope, but rather the specimen itself is scanned by a focused spot of light, and the result of the interaction of the light with successive areas of the specimen is recorded using a non-imaging photodetection device such as a photodiode or photomultiplier tube (PMT). SOM is thus the optical equivalent of scanning electron microscopy (SEM).

In SOM, an objective, usually of high numerical aperture, is used to focus a beam of light to a single diffraction-limited spot within or on the surface of a 3D specimen. This is commonly achieved by allowing the objective to demagnify the light from a laser-illuminated pinhole (the illuminating aperture) into the microscope's focal plane within the specimen (*Figure 1.1b*). We shall term such an instrument a *single-beam* SOM, to distinguish it from alternative designs of instrument, discussed in Section 1.4, in which the object is illuminated by multiple beams formed by apertures in a rotating Nipkow disc, as in the tandem-scanning confocal light microscope, or by a bar of light that is scanned transversely across the specimen, as in the slit-scanning confocal light microscope. In the epi-illumination configuration of a single-beam SOM, the same objective is then used, in conjunction with an appropriate beam splitter or dichroic mirror, to image the reflected light or fluorescence emission from the specimen on to the photodiode or PMT light detector (*Figure 1.2*). Alternatively and less conveniently, the SOM may be used in a transmission mode, in which a second lens is used to image the light on to the detector.

To produce an image using a SOM, as with any single-beam scanning microscope, the illuminating point probe must be moved in a regular 2D raster across or, more accurately for transparent 3D biological prepara-

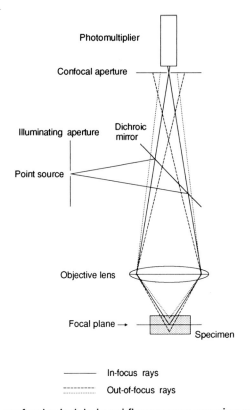

Figure 1.2. The confocal principle in epi-fluorescence scanning optical microscopy. Excitatory laser light from the illuminating aperture passes through an excitation filter (not shown), is reflected by the dichroic mirror and is focused by the microscope objective lens to a diffraction limited spot at the focal plane within the 3D specimen. Fluorescence emissions, excited both within the illuminated in-focus voxel and within the illuminated cones above and below it (shown in *Figure 1.1b*), are collected by the objective and pass through the dichroic mirror and the emission filter (not shown). However, only those emissions from the in-focus voxel (————) are able to pass unimpeded through the confocal detector aperture to be detected by the photomultiplier. Fluorescence emissions from regions below (– – – –) and above the focal plane (· · · · ·) have different primary image plane foci and are thus severely attenuated by the confocal aperture, contributing essentially nothing to the final confocal image. Reproduced by permission of the Royal Microscopical Society from Shotton, D.M. (1988) The current renaissance in light microscopy. II. Blur-free optical sectioning of biological specimens by confocal scanning fluorescence microscopy. *Proc. R. Microscop. Soc.* **23**: 289–297.

tions, *through* the specimen, and the instantaneous response of the light detector at each point in this scan must be displayed with equivalent spatial position and relative brightness on the synchronously scanned phosphor screen of a monitor or on some other suitable imaging device. The image is thus built up serially, point by point, as on the screen of an SEM monitor or of a domestic television, although the scanning rate for SOM is often significantly slower than the 25 or 30 frames per second of conventional video. In contrast to a conventional light microscopic image or its photographic record, the analogue electronic signal produced by the light detector while acquiring a SOM image is easy to digitize 'on the fly' and accumulate in a digital image memory, from which it may be read out and displayed at video rates or subjected to subsequent digital image processing.

1.3 Principle of out-of-focus blur rejection in the confocal fluorescence microscope

In the confocal epi-fluorescence mode of a single beam scanning optical microscope (*Figure 1.2*), fluorescent light is emitted by excited fluorochrome molecules located at the single diffraction-limited point that is illuminated in the focal plane. This light is collected by the same objective, and is brought to a focus via a dichroic mirror and suitable interference filters at an aligned confocal aperture in the primary image plane of the objective. This confocal imaging aperture (hereafter referred to as the confocal aperture) ensures that only light emanating from the in-focus plane is fully detected by the photomultiplier tube (PMT). In contrast, fluorescence emissions excited by the illuminating beam within the conical illuminated regions of the specimen above and below the in-focus plane (*Figure 1b*) come to a focus elsewhere, and so are defocused at the confocal aperture plane, and are thus almost totally prevented from reaching the PMT detector (*Figure 1.2*), contributing essentially nothing to the final image. As a consequence, only in-focus information is recorded. While fewer photons reach the detector than without such a confocal aperture, it is primarily 'noise' (in the form of out-of-focus photons) that is lost, rather than signal. The use of a high-sensitivity PMT enables the confocal technique to be used for fluorescence microscopy in situations where the fluorescence emission intensity of the specimen is only moderate, or where the intensity of fluorescence excitation is intentionally limited to reduce the rate of photobleaching. If the confocal aperture is removed, all of the defocused light is permitted to reach the photodetector, and depth discrimination is thereby abolished; the imaging properties of the non-confocal SOM are formally equivalent to those of a conventional microscope, as described in Chapter 2.

The same basic confocal in-focus selection principle applies when the SOM is operated with a confocal aperture in both transmission mode, and epi-illumination reflection mode (in which a simple beamsplitter is substituted for the dichroic mirror used for epi-fluorescence imaging). However, there is a fundamental distinction between confocal fluorescence imaging and these two latter imaging modes. In the latter cases, the use of an ideally small circular confocal aperture causes the photomultiplier to function as a coherent point detector. Since with laser illumination, the illuminating light is also coherent, the confocal microscope acts as a coherent microscope for transmitted and reflected light. In contrast, because fluorescence emissions are by their nature incoherent, the confocal fluorescence microscope is an incoherent imaging system. This has implications for spatial resolution, discussed in Chapter 2, as incoherent imaging can in principle (according to the particular resolution criterion considered) yield higher axial and lateral spatial resolution.

1.4 Methods of scanning

1.4.1 Single-beam scanning

In practice, single beam scanning can most conveniently be brought about either by the lateral movement of the specimen in the focal plane relative to a stationary optical path (*scanned-stage* SOM), or by the angular movement of the illuminating beam filling the back focal plane of a stationary objective, which causes the focused light beam to move laterally in the focal plane relative to the stationary specimen (*scanned-beam* SOM).

Scanned-stage SOM has the important advantage of constant axial illumination, thus reducing optical aberrations and ensuring complete evenness of optical response across the entire scanned field (space-invariant imaging), a feature desirable for optimal image quality and ease of subsequent image processing. In addition, it permits scanned fields of view much larger than the static field of view of the objective employed, and thus allows one to change between very low and very high magnification imaging while maintaining optimal resolution and light collection, by the simple process of altering the stage scan amplitude while using a single high-magnification objective of high numerical aperture. Such scanning is normally achieved by driving the stage using electromagnetic pushers, such as loudspeaker coils, with scan ranges of up to 5 mm^2 or even greater. However, because of physical limitations of the speed of accurate 2D stage scanning, such scanned-stage microscopes generally have a slow scan rate, usually taking at least a few seconds to collect a high-quality image of at least 512 × 512 pixels (digitized picture elements), although this time may be reduced proportionately by scanning a smaller pixel field. Alternatively, a motorized x, y stage (using d.c. motors or stepper motors) can provide excellent geometric accuracy and even larger scan amplitudes, but is limited in scanning speed to about a minute for a single 2D image. Piezoelectric scanners can be used for high-magnification work, but their scanning range is restricted. Such scanned-stage instruments, primarily used for materials and semiconductor applications, are usually purpose-built electronic imaging devices, and often lack the capability for alternative direct conventional full-field optical microscopic viewing of the specimen through eyepieces.

Scanned-beam SOMs usually use feedback-stabilized vibrating mirrors which can scan at a frequency of up to 1 kHz. They cause angular scanning of an expanded beam of light filling the aperture of the back focal plane of a plan (flat field) objective, allowing higher scanning frame rates, typically between 0.1 and 30 Hz for an image of 512 × 512 pixels. The collected light in an epi-illumination system is then descanned by being reflected by the same scan mirrors before being focused on to the detector aperture. As in conventional optical microscopy, the image field of view in such scanned-beam SOMs, in which the specimen is not moved, is limited to that of the objective lens used, so that a range of objectives is necessary to cover different magnifications. A great practical advantage of this type of SOM for the biologist is that it may be achieved by direct attachment of a scanning unit to a conventional compound microscope, enabling all the normal full-field imaging modes [bright and dark field, phase-contrast, Nomarski differential interference contrast (DIC), conventional epi-fluorescence, reflection contrast, etc.] to be retained and employed on the same specimen. Alternatively, by the combined use of confocal non-immersion microscope objectives and parfocal specialist macro objectives, purpose-built confocal micro/macroscopes can be used to scan in 5 sec fields of view varying from 25 μm^2 to as large as 75 mm^2.

A final alternative method of scanning a single beam across the specimen is that of scanning the objective itself, relative to a stationary specimen and a stationary illuminating system (*scanned-lens* SOM). This method, which retains some of the optical advantages of stage scanning whilst allowing probing of the stationary specimen with electrodes or micropipettes, has been practically implemented by substituting an optical objective for the acoustic lens in a modified scanning acoustic microscope, and has been separately developed as a specialist commercial product for inspection of integrated circuits. However, its main disadvantage is that accurate scanning of a conventional microscope objective is rather slow, and this type of confocal microscope is little used in practice.

1.4.2 Video-rate single-beam SOMs

Recently, various 'real-time' single-beam SOMs have been introduced which give image readout at standard video rates (25 or 30 frames per sec). Such fast-scanning instruments represent a significant advance in technology, and permit, for example, the use of scanning optical microscopy for the study of dynamic physiological processes in living cells and organs. Video-rate scanning gives direct interactive focus control, as in a conventional light microscope. This makes the location of specimens consisting of scattered discrete objects somewhat more convenient, since, when out of focus, such objects are completely invisible when viewed confocally.

8   Confocal Laser Scanning Microscopy

In its simplest and purest form, the video-rate single-beam SOM has been realized by using very rapidly oscillating resonant mirror scanners, retaining true point confocality. An alternative is to use rotating polygon mirror scanners, which achieve very high scanning speeds but which are difficult to synchronize. Rapid beam scanning along the x axis has also

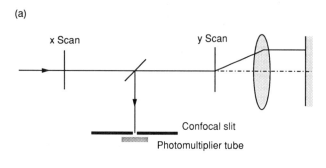

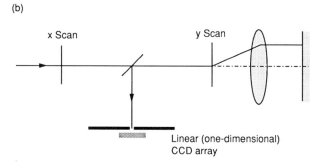

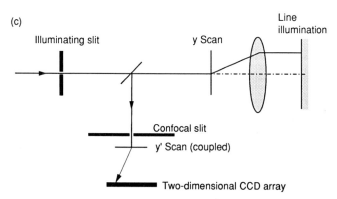

Figure 1.3. Confocal arrangements using confocal slits or line illumination. (a) Conventional 2D scanning spot illumination, using an acousto-optic deflector to achieve a rapid x scan, and a galvanometric y scan. Fluorescence emission is partially descanned by the y scanning mirror, and the partially descanned spot, still vibrating along the x axis, is imaged via a confocal slit on to a photomultiplier tube. (b) Illumination as in (a), but a linear CCD array is used for detection behind the confocal slit. (c) Line illumination using an illuminating slit and an y scanning mirror. A coupled y' scan is used with a confocal slit detector aperture to generate an apparently continuous 2D image that can either be observed directly by eye or imaged using a CCD video camera.

been achieved by the use of an acousto-optic deflector, while using a mirror to scan the beam along the more slowly changing y axis. However, while this arrangement works well for confocal reflection imaging, it presents problems for confocal fluorescence imaging, since the longer wavelength fluorescence emission cannot be descanned back through the wavelength-specific acousto-optic modulator. Instead, the partially descanned fluorescence emission, still oscillating along the x axis, is detected by the photomultiplier through a slit, rather than a circular, confocal aperture (*Figure 1.3a*). The degree to which this slit imaging method compromises the confocal imaging performance is discussed in Chapter 3. Alternatively, the partially descanned spot, still vibrating along the x axis, may be imaged on to a linear CCD detector (*Figure 1.3b*), an arrangement which is useful for metrological work, as described in Chapter 7. Since an image of a single line of the sample is directly recorded by the CCD, geometric distortion is absent and calibration is determined by the dimensions of the CCD. The microscope thus behaves as a conventional microscope for imaging in the x direction, and as a confocal microscope for imaging in the y direction.

All single-beam SOMs suffer from the limitations inherent in serial data collection, namely the necessity to compromise between the rate of image acquisition, the spatial resolution of the raster scan, and the signal-to-noise ratio (*Figure 1.4*). The usefulness of video-rate single-beam scanning systems for fluorescence microscopy is thus still debatable, since at high magnification data acquisition may become photon limited, a low fluorescence signal requiring the subsequent integration of several video frames to achieve a satisfactory signal-to-noise ratio, thereby giving no overall temporal resolution advantage over a slow-scan system.

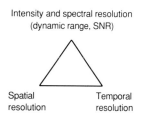

Figure 1.4. The eternal triangle of compromise in designing a single-beam confocal laser scanning microscope. Ideally, one wishes to maximize all three criteria shown. However, the limitations of practical instrument design do not permit this. The nearer one approaches ideal performance in one aspect of data collection, indicated by one of the corners of the triangle, the further one departs from one or both of the other criteria. Reproduced by permission of Springer-Verlag from Shotton, D.M. (1995) Electronic light microscopy – present capabilities and future prospects. *Histochem. Cell Biol.* **104**: 97–137.

1.4.3 Slit scanning

The use of a slit aperture on the illumination as well as the detection pathway, together with a mirror scanning along the y axis only, opens up the possibility of illuminating a whole line of the specimen at a time, using a 2D CCD array camera to capture and transmit the resultant confocal image at video rates (*Figure 1.3c*). Such a microscope thus involves object-plane scanning in the x direction, and image-plane scanning in the y direction. Again, the imaging properties of such systems are discussed in Chapter 3. Use of line illumination (equivalent to a linear array of overlapping points) is an example of parallel data collection by a *multiple beam* SOM, which has the advantage of increasing the optical throughput of the microscope. It is also possible to use other geometries of arrays of points or lines for illumination, for example using a microlens array.

1.4.4 Multiple-beam scanning

One of the first scanning methods used in a practical working confocal microscope employed a rotating Nipkow disc to scan a large 2D array of light points over the specimen, relative to a stationary optical beam, a stationary objective and a stationary specimen. Such an instrument is called a tandem-scanning confocal light microscope. Since, for technical reasons, such confocal microscopes do not usually employ laser illumination, they fall outside the scope of this handbook and will not be discussed in detail.

1.5 Design of single-beam confocal laser scanning microscopes

As is apparent from the preceding discussion, a confocal laser scanning microscope (CLSM) consists of a laser light source, coupled by an intermediate optical system to an objective lens, and then to a photodetector. Most commercial scanning microscopes are scanned-beam systems, designed around a conventional optical microscope, using one or several laser sources (*Table 1.1*). The most popular lasers are the argon ion laser, giving lines at 488 nm and 514 nm, and the argon–krypton mixed gas laser, giving lines at 488 nm, 568 nm and 647 nm. UV excitation is usually achieved by using a more powerful argon ion laser with emissions at 351 and 364 nm, or a helium–cadmium laser emitting light of 325 nm. For illumination in the red or infrared, for example to excite cyanine dyes or to increase penetration through tissue or semiconduct-

Table 1.1. Principle emission lines of gas lasers useful for confocal laser scanning microscopy

Laser	Wavelength (nm)			
	UV	Blue	Green	Red
Helium–cadmium	325	442		
Helium–cadmium (RGB)		442	534, 538	636
Low power argon ion		488	514	
Water-cooled argon ion	351, 364	457, 488	514, 528	
Argon–krypton mixed gas		488	568	647
Helium–neon (green)			543	
Helium–neon (red)				633

RGB, red, green and blue; UV, ultraviolet.

ing material, a low cost helium–neon laser can be used (633 nm or 1152 nm). More powerful infrared pulsed lasers are now being used for two-photon excitation, discussed in Chapter 8.

The radiation from the chosen laser is expanded, using a beam-expander telescope configuration, in order to fill completely the back focal aperture of the condenser or epi-illumination objective lens. This is necessary in order to achieve optimum resolution. *Figure 1.5* shows an objective with finite tube length which requires only a single lens for beam expanding. Usually a spatial filter pinhole (illuminating aperture) (P1) is incorporated in the beam expander in order to produce a uniform illumination beam, although this is not strictly necessary. The aperture size should be chosen to remove the high spatial frequency components of the beam without significantly truncating the Gaussian beam profile. For this reason the optimum size for the illuminating aperture is somewhat larger than that of the imaging aperture required for true confocal operation, discussed in Chapter 3. Many confocal microscopes alterna-

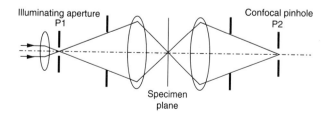

Figure 1.5. The illumination system of a confocal microscope. The radiation from the laser is expanded, using a beam-expander configuration with spatial filter P1, in order to fill completely the back focal plane of the condenser or epi-illumination lens objective.

tively use a single-mode optical fibre to couple the light from the laser to the optical system, thus avoiding vibration problems emanating from the laser cooling fan.

An alternative to a laser source for a single-beam confocal microscope is an arc source, also used in tandem-scanning confocal microscopes. If positioned to give Köhler illumination, the field aperture can be reduced

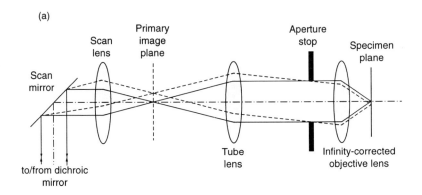

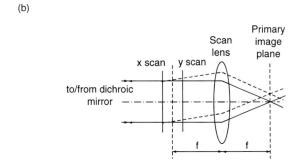

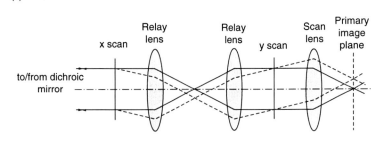

Figure 1.6. The intermediate optical system of a confocal microscope. In order to fill completely the objective pupil (back focal aperture), whilst avoiding shading problems, the scan mirror must be imaged into the primary image plane of the objective. For *xy* scanning a single mirror is sometimes used (a), but if separate mirrors are used they must be either close-coupled (b), or arranged using a telecentric lens relay system (c).

to a pinhole size so that effectively only a single point of the object is illuminated. Obviously the power transmitted through the field aperture under these conditions is small, so this approach is mainly limited to use with highly-reflecting objects such as semiconductors. However, it can be a useful approach because it reduces coherence artifacts in reflection imaging such as speckle and interference fringes.

For scanned-beam operation, in order to fill completely the back focal aperture of the objective while avoiding shading problems, the scan mirror must be placed in the back focal plane of the scan lens, which is conjugate with the back focal plane of the objective. *Figure 1.6a* shows a system with an infinity tube length objective, so that a tube lens is necessary to collimate the light from the primary image plane. This is simple if scanning is required only in a single direction. For xy scanning, the ideal solution is for a single mirror to be used which simultaneously scans in both directions (a cardanic scanning mirror system). Alternatively, separate mirrors may be used provided they are either close-coupled (*Figure 1.6b*) or arranged using a telecentric lens relay system (*Figure 1.6c*). In practice the close-coupled arrangement works adequately and avoids the extra complication and light losses of the relay lens system.

As explained, confocal operation of the microscope is achieved by placing a confocal aperture in front of the photodetector (*Figure 1.2*). The aperture size should preferably be adjustable, either in a continuously variable fashion or by using interchangeable pinholes, in order to obtain the optimum compromise between resolution and signal strength, as discussed in Chapter 3. A single-mode optical fibre can also be used instead of a mechanical aperture. Instead of using two confocal apertures, one for illumination and the other for confocal detection, as illustrated in *Figure 1.2* and employed in most confocal systems, an alternative is to use one aperture only, in which case the dichroic mirror or beam splitter must be relocated to a position between the laser and the aperture (or optical fibre). This arrangement has the important advantage that it eliminates the problem of aligning the two confocal apertures, since the single aperture is automatically aligned. However, reflection from the aperture surround (or optical fibre) results in stray light problems in brightfield reflection imaging, unless special measures are taken.

2 Image formation in the confocal laser scanning microscope

2.1 Principles of image formation in non-confocal and confocal single-point scanning microscopes

The aims of this chapter and the next are two-fold: first to describe the differences between image formation in confocal and conventional light microscope systems and to point out their advantages and limitations; and second to ensure that the microscope user has sufficient knowledge of the image formation process to obtain the best performance from the instrument.

We begin by considering image formation in a conventional brightfield microscope, as illustrated schematically in *Figure 2.1a*. This simplified diagram shows a microscope with critical illumination: a large area incoherent source is focused by a condenser on to the specimen such that a comparatively large area of the specimen is illuminated, corresponding to the whole field of view of the objective (full-field illumination). Information from each illuminated point in the specimen is simultaneously transmitted in parallel by the objective lens to form the primary image. Of course, in a practical microscope, this image is further magnified by the eyepiece. However, the important property to note is that it is the objective which is primarily responsible for forming the image and determining its resolution, with the condenser playing only a secondary role in determining the resolution of the system.

In fact, the aperture of the condenser acts to control the spatial coherence of the overall imaging process, which can vary between fully coherent and fully incoherent. The aberrations of the condenser are not important, so that the source and condenser together behave simply as a partially coherent effective source. For this reason, in the widely used

Köhler illumination arrangement, in which the specimen plane is made conjugate with the field diaphragm rather than with the light source, imaging is in principle identical to critical illumination.

If the numerical aperture of the condenser is made small, compared with that of the objective, imaging becomes coherent. Coherent imaging exhibits fringing on edges in the image, and phase information in the object (which results from optical thickness variations) can be imaged as intensity changes if the system is defocused. Incoherent imaging, on the other hand, is obtained if the numerical aperture of the condenser is made large compared with that of the objective, and although this depends on the form of the specimen, it generally results in superior resolution. It should be noted that it is impossible to obtain truly incoherent brightfield imaging when using a high aperture objective, and that in practice the condenser aperture diaphragm is adjusted to give a compromise between resolution and contrast in the image, resulting in partially coherent imaging. In the case where the numerical apertures of condenser and objective are equal (so-called matched or full illumination), an arrangement that is frequently adopted in practice, imaging, although partially coherent, has similarities with incoherent imaging.

In *Figure 2.1b*, the full-field image of a conventional microscope is measured point by point by a detector in front of which is placed a small aperture. This method is sometimes used for microdensitometry or micrometrology. If the aperture is scanned in a raster, the output of the detector gives a 2D image, in electrical form, which is identical to the

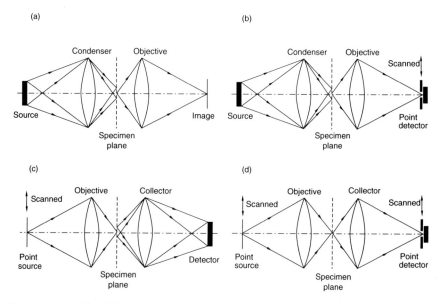

Figure 2.1. Simplified schematic diagram of imaging in transmission microscopes. (a) Conventional optical microscope with critical illumination. (b) As in (a), but the image is measured point by point with a small detector. (c) Single-point scanning optical microscope. (d) Confocal microscope.

optical image observed in *Figure 2.1a*, providing of course that the aperture is small enough that its finite size does not degrade the image resolution, and that the bandwidth of the electrical detection circuit is high enough to transmit the signal variations that occur during scanning. This arrangement is identical in principle to the technique of video microscopy (i.e. image-plane scanning), where the image is scanned off the faceplate of a TV camera tube by a moving electron beam. Because in video microscopy contrast can be electronically enhanced, it is often preferable to use full illumination in that case.

In *Figure 2.1c*, a single-beam object-plane scanning optical microscope is illustrated. A single point in the specimen is illuminated by the focused and demagnified point source, while a large area detector, in conjunction with a collector lens, collects light from the full field of view of the specimen. It will be observed that the ray paths shown in *Figure 2.1b* and *2.1c* are identical but reversed, and, because of the optical principles of reciprocity and equivalence, the microscopes illustrated in *Figures 2.1b* and *2.1c* have identical non-confocal imaging properties. In the scanning optical microscope (*Figure 2.1c*), the resolution of the system is primarily limited by the properties of the first lens, called the objective or projector. The second lens, the collector, controls only the degree of spatial coherence in the imaging process, and its aberrations are unimportant. Because this optical arrangement is analogous to critical illumination, it is called critical detection. An alternative arrangement, in which the detector is placed in the back focal plane of the imaging lens, can be termed Köhler detection: it behaves identically in principle but has the advantage that sensitivity variations in the detector are not so important. Indeed, as the collector and detector together act as an effective detector, the collector may in principle be dispensed with altogether and the detector placed in that plane.

Finally, in *Figure 2.1d*, we combine the arrangements of *Figures 2.1b* and *2.1c* to give a confocal transmission single-beam scanning optical microscope, in which a point source illuminates just a small region of the object, and a confocal point detector simultaneously detects light from this same illuminated region. If the point source and detector are scanned in unison, a two-dimensional image will be generated. However, this imaging system behaves very differently from the previous ones. In fact, it is not possible to devise a non-scanning system which behaves in the same manner: we cannot generate a confocal image by placing a detector or imaging aperture at any plane of a non-scanning system. From the symmetry of *Figure 2.1d* it is clear that the two lenses play an equal part in the imaging process. We might expect that this would result in an improvement in resolution, which is indeed the case. In fact, as we shall describe later, the transmission confocal system behaves as a coherent imaging system, but with a sharper effective point spread function than in a conventional coherent microscope.

Although *Figure 2.1d* is drawn for a transmission geometry, in practice most confocal systems operate in the reflection or epi-illumination mode, in which the same objective lens is used both for illumination and detection (*Figure 1.2*). For a specimen placed in the focal plane, the properties of confocal transmission and reflection systems are identical. However, once the object is moved away from the focal plane, certain differences arise which will be described in Section 2.3. *Figure 2.1d* could equally apply to fluorescence imaging; however in this case, because of the incoherent nature of fluorescence emissions after excitation by coherent light, imaging is always incoherent.

The imaging properties of any lens or imaging system may be described in terms of its 3D intensity point spread function (PSF), $I(x, y, z)$, which is the intensity variation in the image of an ideal point object having finite intensity but no size (or, in practice, of a sub-resolution object such as a 50 nm diameter gold particle or a fluorescent microsphere). While, in conventional fluorescence microscopy the image formation is determined by the PSF of the imaging objective, in a confocal SOM the image formation is determined by the PSFs of both the illuminating and the imaging lenses (objective and collector, respectively in *Figure 2.1d*). This is because both lenses play an equal role in image formation, the former in defining the size of the illuminating spot and the latter in focusing the light emissions from this spot to the confocal aperture. The confocal system PSF is thus the product of the individual PSFs of the two lenses. For a confocal epi-illumination arrangement (*Figure 1.2*), neglecting the change in wavelength of the fluorescent light, these two PSFs are, of course, identical, since the same objective is employed for both illumination and collection. Thus the PSF for an epi-illumination confocal system is equal to the square of the PSF of the objective lens.

2.2 Confocal axial scanning, height profiling and vertical sectioning

The ability of the confocal SOM to record only in-focus information has two further advantages, over and above the ability to collect optical sections. First, it permits the instrument to be operated as though it had an infinite depth of field, rather than an extremely limited one. By slowly and progressively changing the focal plane through the entire 3D specimen (an axial z scan) while performing successive $x\,y$ scans, and by automatically accumulating only in-focus data into a single photographic or digital image throughout this process, a complete in-focus projection of the specimen along the z axis may be recorded as a result of the confocal rejection of out-of-focus information (*Figure 2.2*). Additionally, since the position along the z axis at which the maximum signal is recorded for a particular feature is by definition the in-focus position, axial scan

Image formation in the CLSM 19

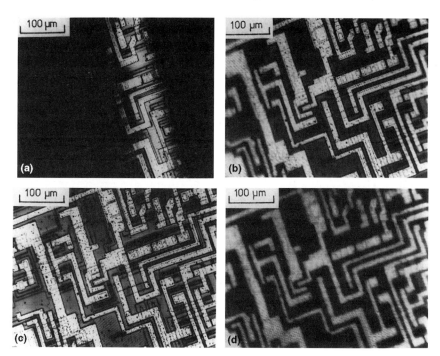

Figure 2.2. (a) A confocal SOM reflection contrast image of a tilted microcircuit, recorded with a 0.5 NA objective using light of 633 nm wavelength, in which only the central in–focus region is imaged strongly, the out-of-focus regions to the left and right, which lie above and below the focal plane, appearing dark. (b) A non-confocal SOM reflection contrast image of the same specimen, in which the out-of-focus left and right regions of the specimen appear blurred. (c) As (a), a 20μm axial z scan of the entire tilted microcircuit through the focal plane of the microscope having been performed during imaging using a piezoelectric z translator, resulting in the accumulation of an in-focus image of the entire specimen. (d) As (c), but with non-confocal optics, the axial scan having resulted in an entirely blurred image. Bar = 100 μm. Reproduced by permission of the Royal Microscopical Society from Wilson, T. and Hamilton, D.K. (1982) Dynamic focusing in the confocal scanning microscope. *J. Microsc.* **128**: 139–143.

data may be used to produce accurate height information of the specimen, which may be displayed in a variety of ways (*Figure 2.3*). Height variations of the order of 1 nm have been clearly distinguished by this method, as discussed further in Chapter 7.

For biologists, one of the most novel and potentially useful imaging modes of the SOM is that of vertical sectioning. Instead of scanning in the $x\,y$ plane, it is quite feasible, while retaining all the advantages of confocal microscopy, to scan in x and z, thus generating images of optical sections parallel to the optical axis of the microscope. Thus transverse optical sections may be chosen of elongated structures (for instance muscle fibres or embryos) lying on a microscope slide with their long axes oriented perpendicular to the microscope's optical axis, and both horizontal and vertical optical sections may be obtained from vertically polarized objects, such as differentiated epithelial cells. Scanning in z may be achieved by a motor drive on the microscope's focus knob, or

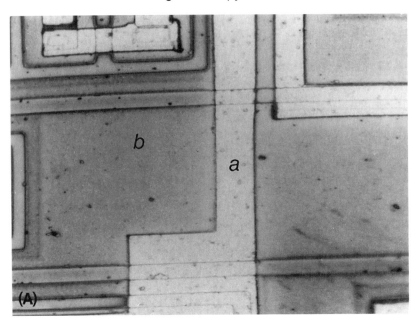

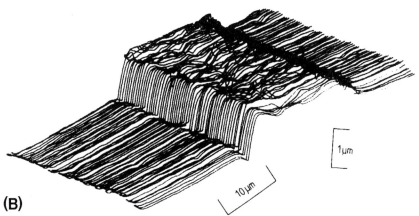

Figure 2.3. A: A non-confocal reflection contrast SOM image of an area of a microcircuit showing an evaporated metal strip (a) about 20 μm wide and 1 μm thick overlying the semiconductor (b). B: Height profile of a section of the metal strip on the semiconductor, determined from the in–focus maxima obtained by confocal axial scanning. Reproduced by permission of Springer-Verlag from Hamilton, D.K. and Wilson, T. (1982) Three-dimensional surface measurement using the confocal scanning microscope. *Appl. Phys. B* **27**: 211–213.

by using a piezoelectric z translator between the microscope stage and the specimen. In principle, by appropriately scanning $x\,y$ and z simultaneously, it is possible to obtain oblique sections at any desired angle. Alternatively, $x\,z$ or oblique sections or projections may be obtained by the subsequent resampling of a complete digital $x\,y\,z$ image, obtained by collecting serial $x\,y$ sections at incrementally increasing values of z, as shown in *Figure 2.4*.

Image formation in the CLSM 21

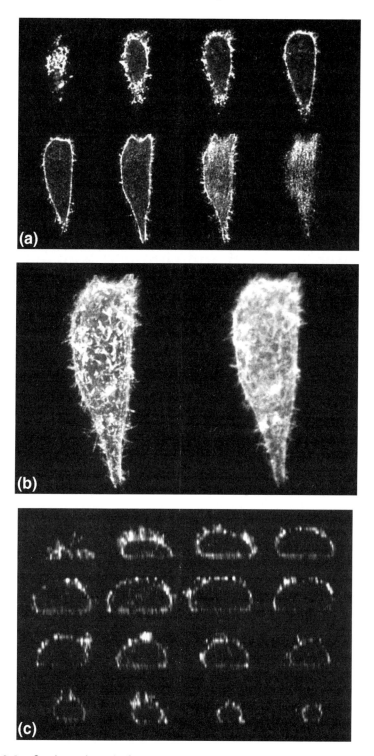

Figure 2.4. See legend overleaf.

Figure 2.4. Confocal fluorescent images of a cultured fibroblast (V79 strain from Chinese hamster lung), about 50 μm in length, in which the F-actin has been stained with rhodamine–phalloidin. A: Eight optical sections selected from the complete data set of 64 sections, displaying (from left to right and top to bottom) the distribution of the cortical meshwork of actin filaments near the top of the cell, around its periphery, and on its ventral surface just above the coverslip on which it was growing. (b) Maximum (left) and average (right) projections of the 3D data set down the z axis. (c) Sixteen computed x z vertical sections at equal intervals from anterior to posterior of the cell, showing the flattened ventral surface adjacent to the coverslip, and actin–rich projections from the dorsal surface. Reproduced by permission of Elsevier Trends Journals and of Springer-Verlag, respectively, from Shotton, D.M. and White, N.S. (1989) Confocal scanning microscopy: three-dimensional biological imaging. *Trends Biochem. Sci.* **14**: 435–439 (selected images), and from Shotton, D.M. (1995) Electron light microscopy – present capabilities and future prospects. *Histochem. Cell Biol.* **104**: 97–137.

2.3 Other imaging modes, including non-confocal and confocal transmission microscopy, confocal reflection DIC, darkfield, polarization and interference microscopy, and optical-beam-induced current (OBIC) imaging

By far the majority of confocal microscopy is performed in the fluorescence or reflection modes, although many commercial systems are additionally equipped with a non-confocal transmission detector that usefully permits the acquisition of a conventional SOM image with whatever optics the microscope is equipped (brightfield, phase contrast, DIC, darkfield, polarization, etc.), while simultaneously acquiring CSOM epi-fluorescence images at one, two or more wavelengths. It is interesting to note that although some of the early work on confocal microscopy was performed in the confocal transmission mode, at present this is not a popular technique. The first reason for this is that, despite the theoretical predictions given in the preceding section, the 3D imaging properties in transmission are not so good. For a weakly scattering object, because of the strong unscattered light in transmission, it turns out that the 3D PSF is not significantly improved, but is almost identical with non-confocal imaging. Thus there is no improvement in 3D imaging performance for such objects. Only for strongly scattering objects is the confocal improvement seen. One potential application for confocal transmission microscopy is the imaging of detail immersed in a highly scattering medium. A major source of contrast in confocal transmission microscopy is caused by an effect analogous to the optical sectioning property of reflection confocal microscopy: the refraction due to the refractive index of the sample produces an effective alteration of the axial position of the confocal aperture proportional to the sample's thickness, thus resulting in a drop in signal. Hence the imaging properties of confocal reflection and confocal transmission are fundamentally different. In particular, although a featureless transverse plane can be located axially in reflection, it cannot be located by confocal transmission. Thus axial imaging in confocal transmission only occurs in association with transverse imaging.

In the same way that refractive effects of the specimen can cause an axial displacement of the focused spot on the aperture, they can also produce a transverse displacement, which is equivalent to a misalignment of the confocal aperture. A number of different ways of overcoming this problem have been proposed, including real-time selection of the optimum aperture position. An alternative method is to use double pass optics, in which the light from the specimen is collected by an objective lens and is then reflected back by a mirror through the objective lens and through the sample. This has the effect of cancelling out the transverse displacement. The light is then detected using an ordinary confocal reflection system, which has the added advantage that the scan mirrors can then be used to descan the beam. Two commercial manufacturers offer systems based on this approach. An alternative possibility is to route the transmitted light back to the illuminating scan mirrors without it passing through the specimen a second time. It should be noted that, if one does not use either of these methods for confocal transmission microscopy, two sets of confocal scan mirrors must be scanned synchronously, one set for the initial illumination scanning and the other for the imaging descanning, which is extremely difficult to do in practice. A simpler alternative, of course, is to use stage scanning, moving the specimen in the x and y planes through the stationary aligned optical beam.

In addition to the straightforward confocal transmission, reflection and epi-fluorescence imaging modes, and the possibilities of non-confocal transmission imaging discussed above, the SOM offers other confocal imaging modes: those of reflection DIC, darkfield, polarization and interference microscopy, and optical-beam-induced current (OBIC) imaging.

Confocal DIC microscopy is a method for observing refractive index changes (phase structure) in a specimen. In the transmission mode, it brings all the problems of confocal transmission imaging. However it is straightforward to perform in reflection. Darkfield confocal microscopy is achieved by using an annular or half-plane obstruction in the illumination path. It can be used to observe very weakly scattering structures, but, as in the conventional darkfield technique, interpretation of the images can be difficult. Confocal polarization microscopy has been performed in both transmission and reflection modes. It has the property that extinction is dramatically improved, so that polarization rectifiers are unnecessary. Confocal interference microscopy has been reported in both the transmission and reflection geometries. Again the transmission mode has various problems. In reflection, a Linnik interferometer is usually used, which can give phase information, or can profile a surface with high sensitivity. The usefulness of many of these alternative imaging modes for biological studies has yet to be fully explored.

In OBIC imaging, the focused laser spot is used to generate electrical carriers in the specimen and the resultant current is monitored. In this way the electrical properties of the sample can be imaged, as discussed more fully in Chapter 7. It has been primarily used with semiconductor materials and devices, but there is the potential for other applications.

2.4 Photodetectors and image noise

2.4.1 Theoretical considerations of photodetector noise

The most important properties of ideal photodetectors are linearity, dynamic range and sensitivity. For fluorescence imaging, where signal levels are frequently very low, the sensitivity and the related low-light-level noise behaviour of the detector are of greatest practical importance. The noise performance when detecting weak signals depends on two fundamental properties of the detector: the quantum efficiency η, defined as the mean number of photoelectrons produced by a single arriving photon (maximally 1.0), and the sensor noise N_e contributed to the final image, measured in electrons per pixel. *Table 2.1* summarizes typical values for various types of detector, some of which are not currently used for confocal microscopy. An ideal detector has a quantum efficiency of one, and zero sensor noise.

The main rival detectors for confocal microscopy at present are the PMT and the CCD detector. The PMT, which is a photometric rather than an imaging device, consists of a photocathode and a series of dynodes by which the number of photoelectrons is multiplied. The quantum efficiency of a PMT is around 13% when equipped with an S20 photocathode, it has good blue sensitivity, falling somewhat for red photons, and because of its high gain, has negligible noise. In contrast, the CCD detector, since it is composed of an array of imaging elements or pixels, is an imaging device. It has a higher quantum efficiency, which is typically in the range 30–50%, but can be in excess of 80% using thinned back-illuminated CCD arrays. When cooled to reduce the dark current noise to negligible proportions, the CCD has only a small but finite sensor read-out noise. It is most sensitive at the red end of the visible spectrum and into the infrared, and is only weakly sensitive to blue and ultraviolet (UV) photons, because of their lower penetration into silicon, unless it has been coated with a special fluorescent coating which absorbs these photons and re-emits them at a longer (yellow) wavelength.

Table 2.1. Performance values for various detectors at 520 nm

Detector	Quantum efficiency η (%)	Sensor noise N_e (electrons/pixel)
Ideal	100	0
Photomultiplier, S20	13	0
Photomultiplier, GaAs NEA	32	0
Cooled PIN photodiode	83	100
Avalanche photodiode	70	0
Cooled frame-transfer CCD	50	9
Interline CCD	14	36
MOS Si sensor	40	745
Si vidicon	90	1500

An advantage of the CCD camera over an intensifying video camera for low-light-level imaging is that it can integrate the signal in analogue form on the chip during a prolonged exposure, which avoids problems with digitization errors and read-out noise encountered when accumulating an image by averaging successive noisy video frames in digital memory. It can also act as a scan converter: the image can be scanned into the device and read out from it at different rates, as happens in slit-scanning confocal microscopes fitted with video-rate CCD cameras. It should be realized that the use of an intensified CCD detector (in contrast to the cooled CCD array detector) involves the use of a photocathode (usually of the S20 variety) in the image intensifier itself, so that although the intensifier gain is higher, its initial quantum efficiency is reduced to that of a photomultiplier tube.

The question for confocal imaging is how the overall performance of PMTs and CCDs compares. The combined effects of quantum efficiency and noise can be examined by considering the signal-to-noise ratio in detection of N_p photons arriving at the detector. The signal-to-noise ratio (SNR) is given by:

$$SNR = \frac{\eta N_p}{\sqrt{\eta N_p + N_e^2}}.$$

Thus at high illumination levels the SNR is proportional to the square root of the quantum efficiency, whereas for a very small number of illuminating photons it is proportional to the ratio of the quantum efficiency to the sensor noise.

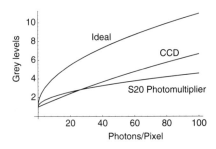

Figure 2.5. The number of grey levels observable with different types of detector at low levels of illumination. For very low light levels, the photomultiplier performs better that the CCD, but for higher light levels the CCD is superior.

The information content, b, in the image, in bits per pixel, is given by:

$b = \log_2 (1 + SNR)$.

In terms of number of significant grey levels, g, into which the digital image can be divided, we can write:

$b = \log_2 g$

so that the number of significant grey levels is simply:

$g = 1 + SNR$.

Note that this derivation of the number of significant grey levels is based on the information content in the image, which is an objective property, rather than on the perceived number of grey levels subjectively discernible by an observer (which is limited to about 30) or the number of digital grey levels into which the image has, for computational convenience, actually been digitized (typically 256).

The number of significant grey levels generated by PMTs and CCDs at very low illumination levels is illustrated in *Figure 2.5*. It can be seen that, comparing the behaviour of a cooled CCD array detector and a photomultiplier tube with S20 photocathode, the CCD exhibits superior performance for exposures of greater than approximately 25 photons per pixel, whereas for less than 25 photons per pixel the photomultiplier tube is better, since the CCD has a low but finite read-out noise at all photon levels. However, in the region where the photomultiplier tube is superior to the cooled CCD, the number of grey levels is less than three, so the image is still of poor quality.

From *Table 2.1*, it can be seen that an alternative photometric device, the avalanche photodiode, combines high quantum efficiency and low sensor noise. However, the device exhibits a long dead-time before another photon can be detected. Thus detectable photon arrival rates are severely limited, and these detectors are at present only suitable for special low-light-level purposes, such as confocal Raman imaging.

2.4.2 Confocal photodetectors in practice

A range of different photodetectors are used for confocal microscopy; while a simple photodiode can be employed for high-light-level reflection or transmission imaging, in most single-beam confocal microscopes one or more PMTs are used. At low light levels, a PMT can be operated in a photon counting mode, which allows discrimination between the arrival of one (or more) photons and dark current (or cosmic ray) noise pulses. Usually, however, the current from the tube is simply measured as an analogue signal using a transresistance amplifier. As the output of the tube consists of a series of pulses, care must be taken when digitizing to ensure that the pulses are adequately sampled: this is conveniently achieved by restricting the bandwidth or by using an integrate-and-hold circuit. Dark current can be reduced by cooling the photomultiplier, although this is rarely done on commercial instruments.

Photomultiplier tubes have either a side-on or an end-on geometry. In the latter, since the photoelectrons must traverse the photocathode, this must be thin with the result that the proportion of photons detected is lower than in the side-on type. On the other hand, the side-on tube generally has a lower gain. One confocal manufacturer employs special optics which essentially combines the best of both designs. Since the photocathode area required for confocal imaging is usually very small, the photocathode size should be minimized to reduce dark current.

As discussed in Chapter 1, charge-coupled device (CCD) detectors are typically used in video rate slit-scanning confocal systems, in which a complete line of the specimen is illuminated at a time and the resulting 2D image is recorded by sweeping the image lines across a 2D CCD array of a CCD video camera. As also mentioned, one or two confocal microscopes use a linear CCD array detector.

2.5 Choice and design of objectives, and avoidance of aberrations

As will be explained in Chapter 3, lateral and, more critically, axial confocal resolution are maximized by the use of an objective with a high numerical aperture (NA). While scanned-stage confocal microscopy, employing on-axis space-invariant imaging, does not require a flat-field (plan) objective, scanned-beam imaging does. For beam scanning systems, a high degree of aberration correction is necessary to obtain good imaging performance, so that plan-apochromatic objectives are preferred. Individual objectives, even of the same type, differ considerably in their optical characteristics and aberrations. These may conveniently be evaluated by determining the axial response of the lens when imaging a plane front-surfaced mirror with confocal reflection optics, making it possible to choose the best objectives for confocal work. This technique may also be used to optimize the imaging performance of the system and compensate for the aberrations introduced by the specimen itself.

It is important to realize that high NA oil-immersion objectives are corrected for observation of specimens close beneath the coverslip (*Figure 2.6a*). Thus when used for optical sectioning deep within aqueous specimens, their performance will become severely compromised by spherical aberration (*Figure 2.6b*). It is very much better to undertake serial optical sectioning of such specimens using a high NA water-immersion objective, either without a coverslip, or with a coverslip if a properly corrected objective is available (*Figure 2.6c*). Use of an oil-immersion objective with a correction collar, which can be used to correct for the spherical aberration, is a good second best if a water-immersion objective is not available.

Absence of chromatic aberration is also important for good confocal fluorescence imaging, particularly for off-axis imaging in scanned-beam confocal systems, since its presence will cause a wavelength- and scan-position-dependent axial shift in the optimal position of the confocal aperture. In practice, this has two consequences. Firstly, it means that if the confocal illuminating and imaging apertures are aligned for a single wavelength and an objective with chromatic aberration is employed, fluorescence excitations induced by different illumination wavelengths will occur at different depths within a double-labelled specimen (*Figure 2.7a*), and thus will not be directly comparable. Secondly, because longer wavelength fluorescence emissions from the illuminated focal plane will not be correctly focused at the confocal aperture (*Figure 2.7b*), the signal strength of these emissions will be reduced.

In the early days of confocal microscopy, these factors presented major problems. However, an increasing number of new (and very expensive) high magnification, high NA water immersion plan apochromat objectives designed specifically for confocal microscopy are now available from microscope manufacturers, which are chromatically corrected throughout the visible or throughout the visible and ultraviolet wavelengths.

Other useful objectives, presently unavailable, would be long working distance ones with high NA and low magnification, since these would retain high light collection, and good axial resolution and confocal stray light rejection when used to observe large fields of view at low magnifications. For brightfield work, objectives specifically corrected for the most frequently used laser wavelength would be advantageous. For scanned stage or scanned lens confocal systems, off-axis aberrations are not important, and hence it is possible to design lenses with improved overall axial performance, such as a combination of high NA and long working distance, at the expense of off-axis performance. As a simple alternative, non-plan lenses can be used, which have fewer optical elements and hence greater optical throughput.

Image formation in the CLSM 29

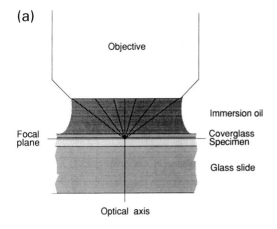

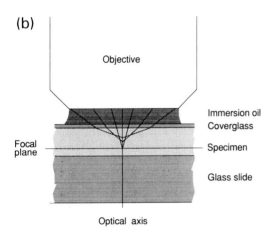

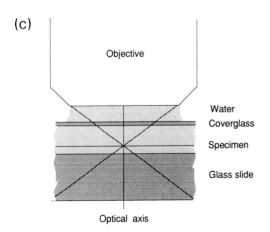

Figure 2.6. (a) Diagram showing imaging by an oil-immersion objective of a specimen located immediately below the coverslip. No spherical aberration is experienced. (b) Ray paths do not converge to a single focus when the same oil-immersion objective is focused deep into the aqueous specimen, because of refraction at the coverslip–specimen interface, which is more extreme for the off–axis rays. (c) With a water-immersion objective, spherical aberration is not experienced when focusing deep into an aqueous specimen. (For simplicity, the small zig-zag refractions occurring as the rays pass through the coverslip, from the immersion water to the aqueous specimen, are not shown). Reproduced by permission of Springer-Verlag from Shotton, D.M. (1995) Electron light microscopy – present capabilities and future prospects. *Histochem. Cell Biol.* **104**: 97–137.

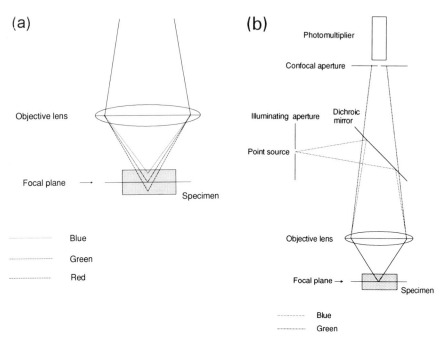

Figure 2.7. (a) When illuminating laser light is focused into a specimen using an objective lens with chromatic aberration, the depth of the focal plane will vary with the wavelength, shorter wavelengths being refracted more and hence focused in a higher plane. The effect is here exaggerated for clarity. (b) With such a lens showing chromatic aberration, in a confocal microscope in which the illuminating and imaging apertures have been made parfocal for the illuminating wavelength, the longer wavelength fluorescence emissions will be defocused at the confocal aperture, leading to loss of fluorescence intensity. Reproduced by permission of Springer-Verlag from Shotton, D.M. (1995) Electron light microscopy – present capabilities and future prospects. *Histochem. and Cell Biol.* **104**: 97–137.

2.6 Setting up and aligning a confocal microscope

Some attention needs be paid to choosing the best location for your scanning microscope. Since vibration can be a problem, the microscope should be mounted on a sturdy table, or even better, a proper vibration isolation table. Fan-cooled lasers can generate strong vibrations of their own, so the fan should be decoupled from the laser head (and preferably situated outside the room, connected with wide diameter ducting) and/or, preferably, the laser decoupled from the microscope and a fibreoptic cable used to bring in the light. Air currents can sometimes cause problems, so try to avoid siting the microscope close to an air-conditioner outlet. It is useful to be able to operate the microscope in the dark, so the room should be made light tight, but dim localized lighting for note-taking will probably be advantageous. Check that an adequate number of electrical outlets and all other necessary services are available; a water-cooled laser will almost certainly require a three-phase electrical outlet in addition to cooling water and a drain (or a coolant recycling system).

When using a modern commercial confocal microscope, there are few alignments which can or need to be made. The most important is the lateral alignment of the confocal aperture(s). A misaligned aperture can reduce brightness or give images with strong fringes, or even a double-image. The best way to align the aperture is to use a mirror as specimen in the reflection mode, or for fluorescence, to observe a thick layer of fluorescent dye or uranyl glass, or fluorescently labelled beads. Then, when scanning in the z direction to observe the axial response, maximize the peak signal by adjustment of the x and y position of the confocal aperture(s).

Axial alignment of the confocal aperture(s) is not usually necessary or possible, although misalignment can again result in image fringing. The most reliable way of checking the axial alignment is to ensure that the focus position remains constant when the illuminating objective aperture is reduced. One situation in which axial alignment is a particular problem is in the use of UV illumination, as a result of the often incomplete chromatic aberration correction of the objective. It is thus common to use an additional correction lens in the UV illumination pathway to compensate for this aberration.

The position of the scan mirror may need to be adjusted in order to avoid vignetting (obstruction of peripheral light rays) in the system. This is a potential problem with systems in which a confocal scanner head must be positioned relative to a conventional optical microscope. In any optical system there is an aperture stop which defines the overall aperture of the system. This is usually within the objective lens but, especially for low magnification lenses, there may be a second aperture elsewhere in the system. If the effective aperture stop of the system is not located in the correct position, the beam incident on the specimen will not simply translate during scanning, but will tend to rotate as well. As a result, the imaging properties will vary across the field, and in particular the off-set illumination will result in differential phase imaging in the brightfield mode.

3 Performance of confocal microscopes

3.1 Confocal optical sectioning and enhancement of axial resolution

There are various ways to quantify the depth discrimination properties of the scanning optical microscope, and to estimate the improvement in axial resolution brought about by confocal imaging, in comparison with a conventional microscope or (which is optically equivalent) with a SOM operated without a confocal aperture. These are based on modelling the imaging of a variety of theoretical object forms (points, lines, planes and extended structures) and on experimental determinations using actual test specimens.

3.1.1 Use of the theoretical PSF to calculate the axial image intensity of a point object

One measure of axial resolution is the variation in the axial image intensity of an idealized point object situated on the optic (z) axis, as the focus of the microscope is adjusted to move the object through the focal plane. Using the expressions given by Born and Wolf (1989) (see Further reading) to define the PSF of the objective lens in a confocal microscope with an ideally small confocal aperture, the axial intensity of the image of such an ideal point object can be calculated at differing degrees of defocus. From this it can be shown that the depth of field of a confocal microscope, defined as the half-width of this variation (i.e. the full width at the half-maximum height, FWHM, of the curve of intensity plotted against axial distance) is reduced relative to that of a non-confocal microscope by a small but significant factor of about 1.4. Such an axial image of a point object is the same in the fluorescence, reflection or transmission modes, assuming that for such model calculations the Stokes' shift is negligible in fluorescence mode.

3.1.2 Spatial frequency cutoff

The most fundamental measure of resolution is the *spatial frequency cutoff* of the imaging system's optical transfer function (OTF). The optical transfer function, being the Fourier transform of the PSF, shows the capability of the imaging system to transmit different spatial frequencies of image information. Based on this criterion, and the fact that fluorescence emissions are incoherent relative to the illuminating light, the maximum axial (and lateral) spatial frequencies transmitted by confocal fluorescence imaging should be twice those of a conventional incoherent fluorescence microscope, again assuming a Stokes' shift of zero. However, it has been shown, both theoretically and experimentally (see Section 3.1.3 below), that the axial resolution of a planar fluorescent specimen is in fact significantly poorer in the case of confocal fluorescence imaging than in the case of confocal reflection imaging, even when assuming the theoretical 'best case' in which the Stokes' shift is zero. The reason for this apparent contradiction is that, although the spatial frequency cutoff of the confocal fluorescence system is very high, the value of the OTF at these high spatial frequencies (i.e. the strength with which the fine detail is actually imaged) is very small. Thus, although the confocal fluorescence microscope has the greatest available resolution in both axial and lateral directions, this is not fully exploited at present in practical confocal microscopes.

3.1.3 Integrated intensity, and axial imaging of points and planes

There are other definitions of depth of field which may be of greater practical importance in fluorescence microscopy. One of these is the *variation of the integrated intensity* of the image of a point object, which is a measure of the total power in its image. This indicates how well the microscope discriminates against parts of a specimen that are not in the focal plane.

In conventional microscopy, the integrated intensity of the light emanating from any one point in the specimen is, to a first approximation, unchanged as one moves away from focus, the light merely being redistributed in the final image and eventually contributing to a uniform background intensity which reduces specimen contrast. This follows directly from the principle of conservation of energy, and explains why the conventional optical microscope has poor resolving power in the direction of the optical axis. In contrast, the use of a confocal aperture causes the integrated light intensity from a single point to fall off sharply as one moves out of focus. For a confocal microscope with an oil-immersion objective of NA = 1.4 and an ideally small confocal aperture, this value drops to about half the original value at a distance of 0.7 λ from the focal plane along the optical (z) axis. In practice, this condition is rarely met, particularly for fluorescence imaging where the size of the confocal aperture is often increased to allow more light to reach the detector, the degree of out-of-focus blur rejection diminishing as larger and larger confocal apertures are employed.

Similarly the conventional microscope is unable to provide any information as to the z position of a specimen consisting of a reflecting plane, or a uniform layer of fluorescing material lacking position-dependent structure in the plane normal to the optical axis (such as a homogeneous solution of a fluorochrome held between a slide and a coverslip). A specimen of this form gives the greatest improvement when imaged in confocal microscopy, and structures of this type provide useful test objects and are commonly encountered in the reflection imaging of microelectronic devices or thin films.

Figure 3.1 shows the theoretical variation in image intensity along the optical axis for a point object and a planar object in confocal reflection and in confocal fluorescence imaging, assuming low numerical aperture and a Stokes' shift of zero. The axial image of a point is the same for confocal reflection and confocal fluorescence. The axial coordinate, u, is normalized and is related to the true z distance by:

$$u = (8\pi n z/\lambda) \sin^2(\alpha/2)$$

where n is the refractive index of the immersion medium, λ is the wavelength of illuminating light and α is the angular semi-aperture of the objective lens (the numerical aperture being defined as $n \sin \alpha$). The advantage of using normalized coordinates is that *Figure 3.1* applies for any wavelength, numerical aperture and refractive index of immersion

(a) Reflection

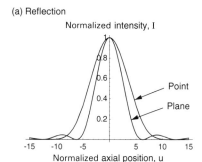

(b) Fluorescence

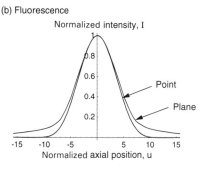

Figure 3.1. Theoretical axial images of a point object and a planar object in confocal reflection and in confocal fluorescence imaging (assuming a Stokes' shift of zero). The image of a point is identical in confocal reflection and fluorescence. The image of the plane is significantly sharper than that of the point in confocal reflection, but somewhat less sharp in confocal fluorescence imaging.

medium, except that some departure from these curves is to be expected at high apertures when the assumed approximations break down. It is interesting to note that, according to this criterion, the axial image of a planar object is considerably sharper in reflection imaging than in fluorescence imaging: indeed the image of a plane in reflection is sharper than that of a point, while in fluorescence mode the axial image of the plane is slightly less sharp than that of a point. The reason why the image of a plane in confocal fluorescence is comparatively poor is that off-axis points in the plane degrade the image, whereas in reflection the phases of these points are such that they add together to give a sharp axial response. These conclusions have been confirmed by experimental measurements.

3.1.4 Variation of axial resolution with numerical aperture

The confocal axial imaging performance depends very highly on the numerical aperture of the objective lens. This is illustrated in *Figure 3.2*, which shows the theoretical confocal axial response, calculated using a high aperture theory, when viewing a plane in reflection with a dry objective of different numerical apertures. It should be noted that the shape of the curves changes slightly with aperture, the intensity minima in particular being weaker for higher numerical apertures. The half-width of this axial response is plotted against NA in *Figure 3.3*. Interestingly, the axial response is sharper for a dry lens than for an oil-immersion lens of the same numerical aperture, because the angular aperture of the dry lens is then greater.

As the size of the confocal aperture is increased, the width of the axial response from a planar reflective object increases so that its reciprocal, the axial resolution, decreases. This is shown in *Figure 3.4*, where the axial resolution in confocal reflection imaging, calculated using a low numerical aperture theory, has been normalized to unity for small circular confocal aperture sizes. The normalized radius v_d is defined in terms of the true radius r_d by:

$$v_d = (2\pi r_d / \lambda) \sin \alpha.$$

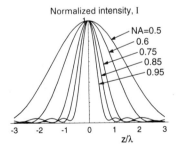

Figure 3.2. The confocal axial response, calculated using a high aperture theory, when viewing a plane in reflection with a dry objective of different numerical apertures (NA). The axial position is given as z/λ.

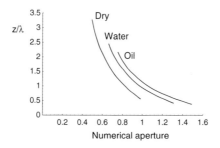

Figure 3.3. The half–width of the axial response from a planar object, calculated using a high aperture theory and expressed as z/λ, plotted against NA for dry, water–immersion and oil–immersion objectives of varying NA. Although higher resolution is achievable using oil immersion, for any particular value of NA a dry objective gives the higher axial resolution.

Similarly, the variations in signal level and axial resolution with the size of the confocal aperture for a planar fluorescent object are shown in *Figure 3.5*. The axial resolution is normalized by that for true confocal reflection. As the diameter of the confocal aperture increases to be equal to that of the Airy disc ($v_d = 3.8$), the signal intensity increases to about 72% of that for a very large aperture, while the axial resolution has only dropped to 81% of that for a true confocal fluorescence system with an infinitely small circular confocal aperture. This diameter thus represents a good compromise between intensity and resolution. The behaviour of a system with a slit aperture of equal width is also shown. The axial resolution with a narrow slit is a factor of 1.4 worse than for a confocal system with a circular aperture of the same width. Although the slit system gives a stronger signal than with a circular aperture of the same width, the axial resolution is thus significantly degraded, so that the slit aperture gives a poorer overall performance. Nevertheless, provided that the slit is not too wide, enough out-of-focus blur rejection is obtained to make this arrangement useful for video-rate confocal microscopes.

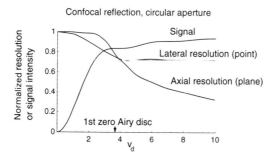

Figure 3.4. The variation in signal intensity and the normalized confocal axial resolution with the size of a circular confocal aperture (normalized to 1.0 for the smallest circular confocal aperture) for a planar reflective object. The size of confocal aperture equal to the first dark ring of the image in the aperture plane is indicated. The lateral resolution of a point object under the same conditions is also shown.

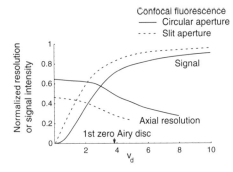

Figure 3.5. The variation in signal intensity and the normalized confocal axial resolution with the size of a circular confocal aperture and of a slit aperture of the same width, for a planar fluorescent object. The axial resolution is normalized by that for true confocal reflection, shown in *Figure 3.4*.

3.1.5 Axial resolution and spherical aberration

As explained in Chapter 2, the axial imaging properties are strongly degraded by the presence of spherical aberration, which can be introduced when focusing with an oil-immersion objective into a thick object of refractive index lower than 1.515 (*Figure 2.6b*). The change in axial response obtained when an oil-immersion objective is used to focus varying normalized distances h inside an aqueous specimen [$h = kn_s d$, where $k = 2\pi/\lambda$, n_s is the refractive index of water (1.33) and d is the true thickness] is illustrated in *Figure 3.6*. The response becomes broad and asymmetric, with pronounced side-lobes, and the peak intensity of the image is also greatly reduced.

Also apparent from these plots is the refractive rescaling of the image: for example, focusing to a normalized depth (kd) of 375 inside an aqueous sample, corresponding to a value of h of 500, results in a peak corresponding to an apparent normalized depth of $kz = 440$. For a wavelength $\lambda = 633$ nm, a layer of true thickness 28.4 μm thus appears to be only 25.2 μm thick.

The ratio of the movement in the focus position to the axial translation of the specimen stage is approximately equal to the ratio of the refractive indices of the specimen and immersion medium, although the exact relationship depends on the numerical aperture of the objective and is, in fact, non-linear with depth. Thus, while 3D data sets collected with oil-immersion objectives need to be corrected geometrically in order to give reliable quantitative axial distances, this cannot be done precisely. It is thus much better to avoid these problems all together by collecting the images using a water-immersion objective.

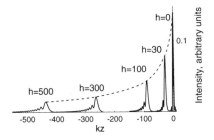

Figure 3.6. The axial response from a planar object obtained in confocal reflection when an oil-immersion objective is used to focus inside an aqueous specimen. As the normalized distance h at which the microscope is focused is increased, the response becomes broader and the peak intensity is reduced.

In summary, it is clear that the optical sectioning performance of a confocal microscope depends crucially upon a variety of experimental variables: the nature and form of the specimen itself; the wavelength of the illuminating light; the Stokes' shift of the fluorochrome employed (in the case of fluorescence imaging); the numerical aperture and degree of aberration correction of the objective; and the size and shape of the confocal aperture. One important conclusion is that significant rejection of out-of-focus information, flare and scattered light will be obtained both with circular apertures significantly larger than required for true confocal imaging, and also with slit apertures (*Figures 3.4* and *3.5*). Another is that it is advantageous to have a confocal microscope equipped with a variable confocal aperture, so that its size may be optimized for each particular application.

3.2 Lateral resolution enhancement

As with axial resolution, the use of an ideally small confocal aperture, which causes the PMT to function as a coherent point detector, results in an improved in-plane image of a point object for confocal reflection imaging (*Figures 3.4* and *3.7*). This is seen as a sharpening of the central peak of the Airy disc with very weak outer rings, which gives a small but significant increase in transverse or lateral spatial resolution (super-resolution) by a factor of approximately 1.4 over that achievable by conventional optical microscopy. The curves in *Figure 3.7*, calculated from the theoretical PSF of the objective lens, are plotted against a normalized transverse coordinate v, which is related to the true transverse displacement r by:

$$v = (2\pi r / \lambda)\, n \sin \alpha.$$

These curves apply for reflection systems and also for fluorescence if the Stokes' shift can be ignored. Together with the improved z resolution discussed above, this modest increase in lateral resolution obtained when using an appropriately small confocal aperture results in an in-focus sam-

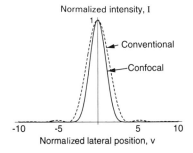

Figure 3.7. The in-plane image of a point object for a confocal scanning optical microscope, showing a sharpening of the central peak of the Airy disc, which gives a small but significant increase in lateral spatial resolution by a factor of approximately 1.4 over that achievable by conventional optical microscopy. The curves hold for reflection systems, and also for fluorescence if the Stokes' shift can be ignored.

pled volume that is about three times smaller than when viewed non-confocally, significantly enhancing the overall optimal performance of the microscope. The 3D image of a single point is thus dramatically improved, as illustrated in *Figure 3.8*. Again, identical results are obtained for fluorescence imaging, if the Stokes' shift can be ignored.

The detector must closely approximate an ideal point detector if such true confocal operation, and the associated improvement in lateral resolution, are to be obtained. The ideally small confocal aperture required to produce this super-resolution permits very little light to reach the photodetector, and thus gives a poor signal level with weakly fluorescing specimens. In practice, apertures larger than that required for true confocal imaging are often employed, in order to increase sensitivity when imaging faint fluorescent specimens, and the benefit of lateral super-resolution is rapidly lost. *Figure 3.4* shows the effect of increasing aperture diameter on the lateral resolution for reflection imaging. The behaviour is identical for fluorescence imaging. If the aperture size is increased to that of the first dark ring of the Airy disc, the lateral resolution improve-

(a)
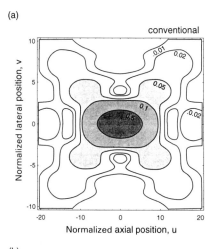

(b)
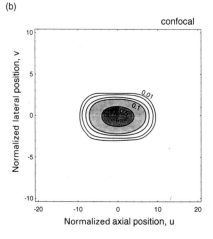

Figure 3.8. The 3D image of a point object, shown as a normalized axial section through the 3D PSF, with (a) conventional and (b) confocal reflection imaging, and also for fluorescence imaging if the Stokes' shift can be ignored. The 3D PSF of the conventional microscope is irregularly shaped, with substantial out-of-focus components, but that of the confocal microscope is compact and well-ordered.

ment disappears. However, axial out-of-focus blur rejection is less sensitive to modest increases in confocal aperture diameter, so that such moderately large apertures may be used to obtain a high signal level while retaining reasonable axial optical sectioning and out-of-focus blur rejection, as discussed in the previous section.

The image of a single point object does not depend on the degree of coherence in the imaging process. Hence in order to investigate the effects of coherence, it is necessary to consider other idealized objects. The imaging of two points is an interesting case. For incoherent imaging, the Rayleigh criterion of resolution is that the points are considered to be just resolved when the distance between the two points, d, is such that the second point is situated over the first dark ring in the Airy disc image of the first point. Under these conditions, it is found that:

$$d = \frac{0.61\lambda}{NA}.$$

This occurs when the ratio of the intensity midway between the points to that at the points themselves is 0.735. This value is adopted as the generalized Rayleigh criterion for the resolution of two points in a general optical system. *Figure 3.9* shows theoretical images of two points at different spacings in various imaging systems. The normalized spacing between the points that corresponds to an actual Rayleigh resolution separation, d, is equal to $2v_0$, where $v_0 = 1.92$. It is seen that for this case the confocal fluorescence image (*Figure 3.9e*) is the best.

The values for the generalized Rayleigh separation $2v_R$, obtained by calculation of the images for different point spacings, are as follows: for conventional coherent imaging (*Figure 3.9a*), $2v_R = 5.16$; for conventional incoherent or full illumination imaging (*Figure 3.9b* and *3.9c*), $2v_R = 2v_0 = 3.84$; for confocal reflection imaging (*Figure 3.9d*), $2v_R = 3.54$; and for confocal fluorescence imaging (*Figure 3.9e*), $2v_R = 2.92$. Thus the confocal fluorescence microscope can resolve two transverse points 1.32 times as close as can be resolved using the conventional fluorescence microscope. Comparing this figure with the value of 1.4 given for the resolution improvement as defined for a single point object, it is seen that the resolution improvement of confocal fluorescence imaging is not quite so large for two points as for a single point.

Confocal imaging of fluorescence emission involves an additional difference compared with conventional imaging. Fluorescence emission generally occurs at a longer wavelength than that of the excitation, by a factor determined by the Stokes' shift of the particular fluorochrome employed. The spatial resolution of confocal fluorescence imaging will thus be determined both by the illuminating wavelength (λ_1) and by the emission wavelength (λ_2), since in the confocal system it is the shorter wavelength excitatory light, λ_1, which determines the size of the illuminated spot. Thus, in confocal fluorescence microscopy, the spatial fre-

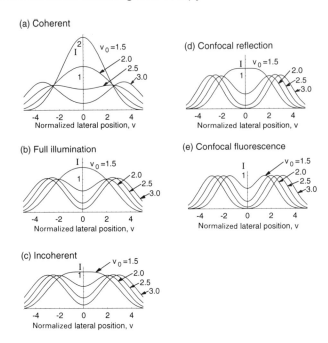

Figure 3.9. Images of two points at different separations in various imaging systems. The distance between the points is described by the normalized parameter v_0, which has a value 1.92 for a separation corresponding to the Rayleigh resolution. It is seen that, for two points, confocal fluorescence results in the largest drop in intensity mid-way between the points, that is the best resolution.

quency cut-off is proportional to $(1/\lambda_1 + 1/\lambda_2)$. This gives the confocal scanning microscope, when used in fluorescence mode, an *additional* resolution advantage over conventional fluorescence microscopy, in which the spatial resolution is determined solely by the longer wavelength, λ_2, of the fluorescence emission. Since resolution decreases with increasing wavelength, choice of fluorochromes with shorter excitation wavelengths and smaller Stokes' shifts will maximize the spatial resolution of the images obtained, although when working with living cells this must be balanced against the phototoxicity of ultraviolet and blue light.

3.3 Reduced image degradation due to light scattering and autofluorescence

In conventional transmission or epi-fluorescence microscopy, the entire observed area of the specimen is uniformly bathed in light (*Figure 1.1a*), but each point of the image only collects a small fraction of the flare, light scattering or fluorescence emissions from the specimen, optical components and the immersion oil. In the single-beam SOM only one point in the focal plane of the specimen is illuminated with full intensity at any one time, illumination of other regions of the specimen being limited to an expanding cone above and below the focal plane, in which the intensity is inversely proportional to the square of the distance from the focal plane (*Figure 1.1b*). As the large area detector of a non-confocal SOM system collects a much larger fraction of the stray light, the overall behaviour is no better than for the conventional one. In fact the behaviour is identical: this is a consequence of the principle of reciprocity. However, when imaging confocally, the rejection of light emanating from above and below the focal plane means that image degradation due to flare, light scattering or fluorescence emissions from other parts of the specimen is greatly reduced, and that autofluorescence contributions cease to be a problem. Thus the image contrast of the CSOM in the fluorescence mode is significantly enhanced over that of a conventional fluorescence microscope.

A convenient measure of this image contrast is the detectability of a point object within background fluorescence, defined as the signal from a point object divided by the noise from a featureless background volume. The variation in 3D detectability, D_{3D}, with normalized confocal aperture diameter, v_d, is shown in *Figure 3.10*. The optimum performance is achieved when $v_d = 2.4$, equal to 0.63 of the radius of the first dark ring of the Airy disc. This emphasizes the fact that the aperture size must be kept quite small in order to achieve optimum image contrast.

Based on the property of 3D detectability, and assuming that the same exposure to light takes place in each case, so that the number of available signal photons is limited by bleaching of the fluorochrome, the performance of various systems can be compared. Systems with line illumination and line detection, and systems using slit detection after point illumination, both give a detectability about one half that of the point scanning confocal microscope with a circular confocal aperture, whereas the conventional microscope is an order of magnitude worse, because out-of-focus blur increases the background signal (*Table 3.1*).

However, for fast scanning confocal fluorescence systems, the fundamental limit to signal is set, not by bleaching, but by the number of

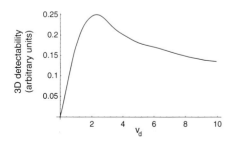

Figure 3.10. The variation in 3D detectability with normalized aperture size. The 3D detectability measures how easily a point fluorescent object can be detected in a medium of uniform background fluorescence. Optimum detectability is achieved for $v_d = 2.4$, corresponding to a confocal aperture about 0.63 times the size of the first dark ring of the Airy disc.

Table 3.1. Relative detectability of a point object within background fluorescence in microscopes of various types

Microscope type	Detectability limited by photobleaching (a)	Detectability limited by saturation (b)
Confocal	1.0	1.0
Line illumination	0.5	11
Slit detection	0.4	0.4
Conventional	0.08	40

The relative detectability of the confocal microscope is taken as unity. Figures are shown for the cases (a) when performance is limited by photobleaching, appropriate when unlimited time is available to produce the image, and (b) when performance is limited by saturation, as in video-rate imaging.

available photons, which reaches a maximum upon saturation of the fluorochrome. Now the relative performance is very different: slit detection following point illumination is still worse than circular aperture detection, but line illumination is now an order of magnitude better, because a whole line of the image is recorded in parallel (*Table 3.1*). This explains the attraction of line illumination systems for video-rate imaging. Conventional imaging is even better, because the whole object is imaged in parallel.

3.4 Sensitivity, geometric linearity, dynamic range and photometric accuracy

Detailed quantitative studies of imaging sensitivity using CSOM are not yet available, but it is clearly more sensitive than direct photography using high-speed film. By using digital image averaging (Chapter 4) to reduce temporally varying noise, confocal fluorescence microscopy can indeed be used for the detection of weak fluorescent signals.

As with cooled CCD array cameras, SOMs in which scanning and image digitization are correctly performed do not suffer from the geometric image non-linearity characteristic of intensifying video cameras. Furthermore, the use of a PMT to measure the incoming light ensures a high dynamic range and good photometric accuracy. Thus CLSM images resemble full-field CCD images in being intrinsically of high data quality, although quantitative comparisons between the two image types have not yet been made. It is thus regrettable that in most commercial confocal microscopes the dynamic range of the recorded confocal images is limited, for computational convenience, to 8 bits (256 grey levels).

4 Image processing of confocal image data sets

4.1 Computational requirements

While it is entirely feasible to operate a scanning optical microscope as an entirely analogue instrument, the suitability of the linear data stream of the single beam SOM for digitization, and the usefulness of subsequent digital image processing and image storage techniques, has resulted in most SOMs being configured as digital imaging systems, interfaced to a computer which is used to control the microscope scanning, image acquisition and display synchronization. Many of the prototype SOMs and a certain number of commercial instruments have been built around dedicated microprocessors. Others, particularly those which can be fitted as accessories to conventional optical microscopes, are interfaced to a standard personal computer (PC) or workstation, usually equipped with proprietary scan control and image processing boards. Such a computer will come with sophisticated menu-driven software, making a compact and easy to use stand-alone system for data acquisition and image display.

In the past, such a system typically incorporated a digital image memory ('frame store') capable of storing several full-sized (e.g. 768 × 512 pixel) monochrome images, in which the intensity of each pixel is encoded as an 8-bit (one byte) integer, representing one of 256 possible grey levels from black to white. More recently, however, as microcomputers have become ever more powerful and well-configured, use of such dedicated image memory has given way to straightforward use of the computer's main random access memory (RAM) and video display capabilities. This memory is used to store the incoming slow-scanned monochrome image from each photodetector channel, and to supply it to a monitor for display. As discussed in Chapter 5, a modern multiparameter confocal system can acquire up to five images synchronously, and can display up to four of these side-by-side on a high-resolution monitor, or combine them into a composite pseudocolour image.

A minimal specification for such a confocal system host computer at the time of writing would be one having a 200 MHz Pentium processor (or equivalent), 32 Mbytes of RAM and a 2 Gbyte hard disc. Nevertheless, despite the growing computational power of the PC, the requirement for sophisticated post-processing of large 3D confocal image data sets may mean that this type of work is best transferred to a more powerful graphics workstation, freeing the confocal microscope's host PC for its primary task of image acquisition.

One important consideration, particularly for 3D imaging, is the amount of digital storage required for confocal images. To represent completely an image of the real continuous world by its digitized samples, one must sample it at the Nyquist sampling frequency of twice the spatial resolution of the optical system. For high resolution confocal microscopy, this involves sampling x and y at approximately 100 nm intervals, and z, because of the intrinsically poorer axial resolution even of the confocal microscope (see Chapter 3), at intervals of around 400 nm. Subsequent image processing, particularly stereoscopic display, must take account of the non-cubic nature of these voxels. At this resolution, a single-wavelength confocal optical section of 768 × 512 pixels will cover an area of approximately 75 µm by 50 µm, and will require 0.375 Mbytes of digital image storage if digitized to 8 bits (256 grey levels). Similarly, a single-wavelength 3D image approximately 75 µm by 50 µm in area and 50 µm deep, comprising 128 $x y$ optical sections separated by intervals of 0.4 µm, would occupy 48 Mbytes. If more than one wavelength is imaged, the storage requirement increases proportionately. For four-dimensional studies (x, y, z and time) of living cells, smaller sub-images must be used, to prevent the storage requirement becoming astronomically large. If, for example, 20 time-points are used and sub-images of only 256 × 256 × 64 voxels are selected (i.e. a 25 µm^3 sampled volume), storage of 80 Mbytes of uncompressed digitized data would be required for each 4D image. PC hard discs of 4 Gbyte capacity are now routinely obtainable, so that initial data storage presents no problem, but to enable one to process these multidimensional images, and to display animations computed from them, without having to wait for disc access, at least 64 Mbytes of RAM is recommended so that the entire image dataset can be held in memory. For long-term image archiving, it is necessary to off-load the digital images to some form of removable medium, such as an optical disc, a writable CD-ROM, or (less satisfactorily) a DAT magnetic tape. Lossless data compression procedures, such as run-length encoding, are obviously desirable for large data sets.

4.2 Use of pseudocolour in multiparameter imaging

By assigning the individual grey level values of an 8-bit monochrome image to particular colour values, monochrome images may be displayed in an infinite variety of *pseudocolours*. Since the eye is better able to discriminate between different shades of colour than between different shades of grey, pseudocolour display is particularly useful for highlighting particular regions of a grey-level image, or for distinguishing between areas of similar grey-level (*Figure 4.1*, Colour Plate I). Pseudocolour is widely employed, for example, to display spatial variations in the concentration of intracellular free calcium ions determined by fluorescence ratio imaging (see Chapter 5). It is also frequently used to display merged monochrome images obtained at different wavelengths by multiparameter confocal fluorescence microscopy of multiply-labelled specimens, in which the fluorescence images recorded for the individual fluorochromes are displayed in the colours naturally emitted by those fluorochromes, for example fluorescein in green and rhodamine in red, thus enabling the combined pseudocolour confocal image to resemble colour photomicrographs made directly from dual-labelled specimens (*Figure 4.2*, Colour Plate I).

The pixel intensity distributions in the two fluorescence channels of double-labelled specimens can be analysed statistically by computing 2D pixel fluorograms, in which the population histograms of both channels are plotted with corresponding colour, those pixels which are equally intense in both channel images appearing with intermediate colour along the diagonal of the pixel fluorogram (*Figure 4.3*, Colour Plate I). This type of analysis has some similarities to the 2D cytofluorograms employed to display the result of flow cytometric analyses of cell populations, except that in the cytofluorograms each spot represents one cell, whereas in these pixel fluorograms each spot represents all those pixels having a particular ratio of the two fluorescence intensities in the original image. Such pixel fluorograms are particularly useful for analysing and correcting image acquisition noise, photobleaching, movement in living specimens, background fluorescence and fluorescence cross-talk between channels.

4.3 Three-dimensional image processing

Since the confocal laser scanning microscope may be used directly for non-invasive serial optical sectioning, yielding high resolution images essentially free from out-of-focus blur, it permits the direct acquisition and subsequent study of complete in-focus 3D data sets.

Once a series of non-invasive optical sections has been collected by CSOM, the 3D data set which these comprise may be manipulated in many ways (see *Figure 2.4*), in common with similar data sets derived from deblurred conventional optical sections, and with those obtained from X-ray or magnetic resonance tomographic imaging of the human body, or electron density maps of protein structures calculated from X-ray crystallographic analyses, to take two familiar non-microscopic examples. A great advantage of CSOM optical sections is that they are already in spatial register, and thus do not require computational alignment before subsequent processing. *Figure 4.4* (Colour Plate I) shows three such confocal sections of a pollen grain, a test specimen widely used to demonstrate confocal sectioning because of its bright autofluorescence and striking morphology, while *Figure 4.5* (Colour Plate II) demonstrates how different views of *Paramecium* may be obtained from a single 3D confocal image by computing projections using either the top five or the central 14 sections from the entire data set.

The availability of high resolution non-overlapping in-register optical section series makes it possible to undertake the accurate quantitation of structural features in the specimen which has been thus sectioned, using a variety of recently-devised unbiased estimators for 3D stereological analysis. In addition, there exists a wide variety of image processing and analysis procedures that can be applied to further enhance the images, or to extract features from the background for subsequent measurement or display, exemplified by the 'seedfill' algorithm used to isolate the three individual Golgi-stained neurones shown in different orientations in *Figure 4.6* (Colour Plate II).

Confocal images, for all their advantages, are not perfect, and it is possible to enhance them further by subsequent computational image restoration processing to remove residual blurring due to the 3D point spread function of the confocal optical system.

4.4 Stereoscopic imaging

Despite the fact that a significant fraction of the population (about 10%) is unable to perceive depth by stereopsis, one of the most straightforward and useful image manipulations which may be applied to 3D confocal data sets is the generation of stereoscopic pairs of images, permitting the results to be inspected directly in three dimensions. It is usual to collect a complete series of confocal optical sections, and to generate the stereo pairs, either 'on the fly' or subsequently, by the mathematical projection of the complete 3D data set down two selected inclined axes corresponding to the required left and right eye views. This may be achieved by combining all the sections, pixel by pixel, after small incremental shifts of successive sections by a constant displacement to right or left. For shifts involving integral numbers of pixels, this is straightforward to achieve, enabling rapid stereoscopic display, although this method places restrictions on the possible stereo angles obtainable between the left and right eye images. Complete flexibility, at the expense of speed, may be obtained by non-integral shifts involving pixel grey-level interpolation, achievable in software or by use of a sophisticated image processor. Surprisingly few sections (as few as 12) are required to generate a stereoscopic image in which the individual planes are not perceivable, reflecting the fact that the human eye is less sensitive to variations in stereoscopic depth than in lateral directions.

Such stereo image pairs may be viewed stereoscopically in a number of ways designed to ensure that the right image is not seen by the left eye and vice versa, the most common of which, in order of increasing sophistication (and cost!), are as follows: The images may be displayed (a) side-by-side, as left-right pairs of photographic prints or images on a single video monitor, requiring parallel or cross-eyed stereoscopic fusion, either by 'free-viewing' or by using a simple optical aid incorporating lenses and/or mirrors (see *Figure 6.1*); (b) as a single red/green anaglyph photograph or image on a colour monitor viewed through red/green spectacles, although this method precludes the use of colour for other purposes (*Figure 4.7*, Colour Plate II); or (c) as full-screen colour images occupying alternate frames on a single high frequency non-interlaced monitor, viewed through liquid crystal shutter spectacles, in which the right and left eyepieces are alternately made opaque in synchrony with the left and right image display.

As shown in *Figure 2.4b*, the manner in which the individual sections should be combined for this purpose is to some extent a matter for subjective evaluation, depending upon the nature of the specimen being imaged. The simplest procedures are merely to average the pixels in each projected column (an 'average' projection), or alternatively to record the value of the brightest pixel along each projection line (a 'maximum' projection), a procedure that works surprisingly well for many specimens, despite its inability to remove hidden lines and surfaces. However, more convincing images may sometimes be obtained by exponentially weighting successive sections and then averaging, such that the front of the specimen is made to appear brighter or in a different pseudocolour than the rear, thereby adding an extra depth cue to that of parallax. Alternatively, if the front-most or rear-most voxel exceeding a given threshold is selected, a 'threshold' projection of the front or rear surface of the image is obtained.

4.5 Surface and volume rendering

An alternative non-stereoscopic image display mode for a 3D CSOM data set, particularly useful for compact fluorescent objects, is that of displaying them as if surface shadowed or surface luminant, by making use of surface rendering procedures that are standard on graphics workstations. Here the primary problem is that of defining the object surface required for these procedures from the original continuous unsegmented 3D voxel image, using some kind of thresholding algorithm. Having identified the voxels defining the surface of the object to be displayed, the other voxels of the image are discarded, to produce a very much smaller data set which can be handled rapidly by a graphics workstation. More sophisticated volume rendering or voxel rendering algorithms, which do not discard non-surface voxels, are preferred for diffuse or extended objects, such as cytoskeletal arrays, which do not readily lend themselves to surface rendering techniques.

By altering the constants that model transparency or reflectivity, the rendered objects can be displayed as opaque or semi-transparent, so that it is possible to see structures within other structures, for instance the nucleus within a cell. It is further possible to use an 'electronic knife' to section the model at any arbitrary angle and look within. While extremely useful for the non-stereoscopic publication of 3D data, such rendered images may themselves be made stereoscopic, heightening their impact.

4.6 Computer animation of 3D and 4D confocal images

The ultimate development of CSOM image display is to compute not just a single image or a stereoscopic image pair, but an incremental series of such pairs that may subsequently be displayed from digital memory in rapid succession, forming an animation sequence, which itself may be stereoscopic. If the successive images or stereo pairs are calculated at small incremental circumferential angles, upon replay the object will appear to rotate before one's eyes (*Figure 4.6*, Colour Plate II). If, alternatively, successive images are calculated at increasing distances along one axis, for instance down the z axis, or alternatively along the y axis for a series of transverse $x\,z$ vertical sections of an elongated structure such as an embryonic neural tube, then upon replay one will have the illusion of moving progressively along the specimen. For this latter approach to work most effectively, perspective rather than orthogonal projections are required, giving the viewer the impression of actually floating through the structure.

The amount of memory required for the storage and display of pre-computed animation sequences may be appreciated by considering the rotational animation of a medium resolution 3D image, each projected image of 512 × 512 pixels being calculated at 5° intervals (72 views). To provide the ability to display simultaneously any two of the three possible orthogonal rotational viewpoints ['tumble' (x), 'roll' (y) and 'spin' (z)], without reading new data from disc, 36 Mbytes of available image display memory would be required.

This technique can be further extended by sequentially displaying a series of animations of the same space which have been recorded at successive times. Time-lapse 4D animation display may be achieved by storing the complete animation sequences of two adjacent timepoints at any one time, the first being overwritten from disc by that for the timepoint-after-next while the second is being displayed. Such a four-dimensional (x, y, z, t) display of a time-lapse series of confocal images collected from a 3D specimen, for example, a living cell undergoing a physiological or developmental process, or a living embryo in which certain cells had been specifically labelled with a fluorescent marker, enables one to study phenomena such as membrane dynamics, or cellular migration during early embryonic development, with a clarity and facility impossible without confocal imaging. Confocal scanning optical microscopy has thus brought us the ability to view stereoscopically, rotate, quantify and transform blur-free images of materials or cellular ultrastructure with an ease akin to that exemplified by earlier advances in the fields of whole body medical imaging on one hand and molecular graphics on the other.

Colour Plate I

Figure 4.1. Confocal fluorescence optical section of a bone marrow-derived dendritic cell cultured in the presence of granulocyte macrophage colony stimulating factor, as described in Winzler, C., Rovere, P., Rescigno, M., Granucci, F., Penna, G., Adorini, L., Zimmermann, V.S., Davoust, J. and Ricciardi-Castagnoli, P. (1997) Maturation stages of mouse dendritic cells in growth factor-dependent long-term cultures. *J. Exp. Med.* **185**: 317–328. The MHC Class II molecules are labelled with a fluorescein-conjugated antibody and displayed using the rainbow pseudocolour intensity scale given at the top left. Dendritic cells have arborescent dendrites and cell surface projections covered with MHC molecules. Total field of view: 100 µm. Courtesy of Patrizia Rovere, Claudia Winzler, Paola Ricciardi-Castagnoli and Jean Davoust (CNR di Farmacologia, Milan, Italy, and Centre d'Immunologie CNRS-INSERM, Marseille, France).

Figure 4.2. A merged pseudocolour confocal projection image of actin microfilaments (red) and microtubules (green) in motile cytotoxic T lymphocytes. After fixation, permeabilization and blocking of the cells, microtubules were immunocytochemically labelled with fluorescein, while F-actin filaments were stained with rhodamine–phalloidin. In the final pseudocolour image, actin appears red and tubulin is shown as green. The cells are approximately 10 µm in diameter. Reproduced by permission of Springer-Verlag from Shotton, D.M. (1995) Electron light microscopy – present capabilities and future prospects. *Histochem. Cell Biol.* **104**: 97–137, where further details are given.

Figure 4.3. (a) Confocal immunofluorescence optical section of double-labelled fibroblast cells. The lysosomal enzyme cathepsin D labelled with fluorescein (green) is found in intracellular vesicles, whereas Texas Red-coupled wheat germ agglutinin (WGA) (red) binds mostly at the cell surface. Field of view = 50 µm. (b) Fluorescence intensity-weighted pixel fluorogram from the double-labelled confocal immunofluorescence micrograph shown in (a) (compensated for cross-talk of fluorescein emissions in the Texas Red channel). Fluorescein and Texas Red pixel values are represented along the x and y axes respectively. Reproduced by permission of the Royal Microscopical Society from Demandolx, D. and Davoust, J. (1997) Multicolour analysis and local image correlation in confocal microscopy. *J. Microsc.* **185**: 21–36, where the procedures and advantages of this method of analysis are detailed.

Figure 4.4. Three optical sections through a pollen grain of *Passiflora coerulea*, showing autofluorescence when excited at 488 nm. Note that information present in one optical section is completely excluded from the other sections by the confocal optics. Images courtesy of Anna Smallcome, Bio-Rad Microscience Ltd, who retains the copyright to this image.

Colour plates kindly sponsored by Bio-Rad Microscience Ltd

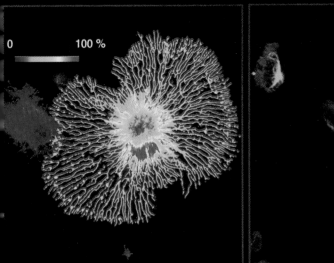

Figure 4.1

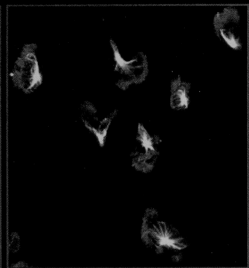

Figure 4.2

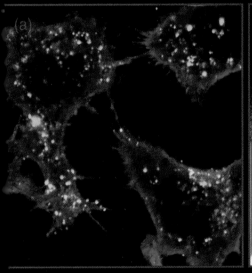

(a)

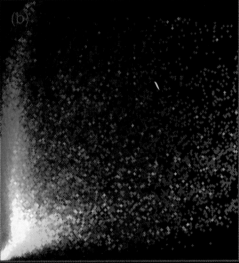

(b)

Figure 4.3

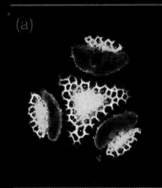

(a)

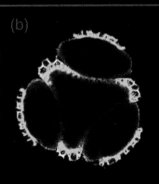

(b)

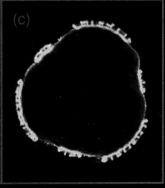

(c)

Figure 4.4

Colour Plate II

Figure 4.5. *Paramecium*, labelled with a fluorescein-conjugated polyclonal anti-axonemal tubulin antibody, showing many hundreds of surface cilia. (a) Maximum brightness projection of sections 22–26, showing upper surface. (b) Maximum projection of sections 8–21, revealing details of internal microtubules associated with the oral cavity and the contractile vacuole. Scale bar = 25 μm. Image (b) first published in Fleury, A. (1991) Microtubule density in ciliated cells: evidence for its generation by post-translational modification in the axonemes of *Paramecium* and quail oviduct cells. *Biol. Cell* **71**: 227–245, and both parts (a and b) previously published in Bio-Rad Confocal Application Note 10 by Bio-Rad Microscience Ltd. Reproduced by permission of Publications Elsevier (b) and of Bio-Rad Microscience Ltd (a and b), and courtesy of Anne Fleury (Laboratory of Cell Biology 4, University Paris-Sud, 91405 Orsay, France).

Figure 4.6. Three viewpoints from a rotational animation sequence showing three Golgi-stained rat brain neurons that have been 'extracted' from a complete 3D confocal reflection contrast image using a 'seed-fill' algorithm, and are here displayed in three separate pseudocolours. Images courtesy of Anna Smallcome and Robert Fairlie, Bio-Rad Microscience Ltd, who retain the copyright to these images.

Figure 4.7. Anaglyph stereoscopic 3D image of the fibroblast shown in *Figure 2.4*, which should be viewed through red (left)/green (right) spectacles. Reproduced by permission of Elsevier Trends Journals from Shotton, D.M. and White, N.S. (1989) Confocal scanning microscopy: three-dimensional biological imaging. *Trends Biochem. Sci.* **14**: 435–439.

Figure 5.1. Monochrome confocal image of well-spread interphase cultured fibroblasts, each about 50 μm in diameter, stained with a fluorescein-conjugated secondary antibody and a primary antibody against tubulin. Image courtesy of Anna Smallcome, Bio-Rad Microscience Ltd, who retains the copyright to this image.

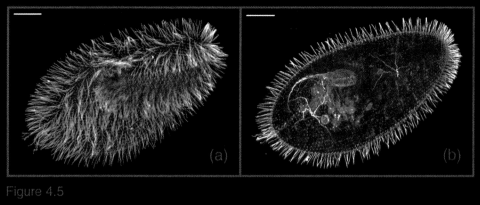

Figure 4.5

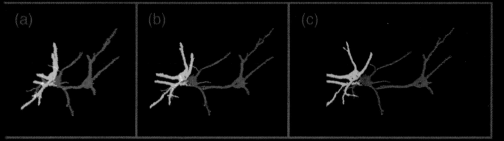

Figure 4.6

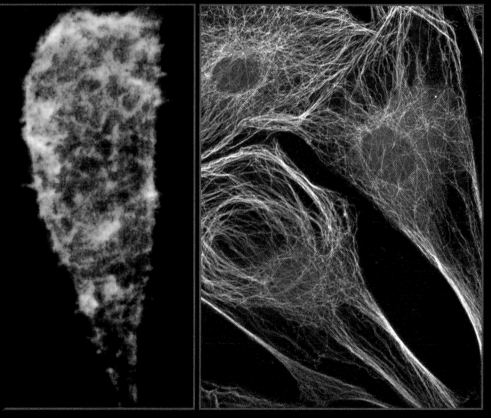

Figure 4.7   Figure 5.1

Colour Plate III

Figure 5.2. Confocal optical section through an intact isolated ventricular cardiac muscle cell stained for α-actinin using a rhodamine immunoconjugate. The cell, which is approximately 25 µm in width, is packed full of contractile myofibrils, seen as repeating striations where the Z-disks, which lie in register, have been labelled by the antibody. This image was prepared by Shirley Stevenson (Imperial College School of Medicine, National Heart and Lung Institute, Royal Brompton Hospital, London SW3 6NP, UK), and won first prize in the Confocal Microscopy category of the Royal Microscopical Society's 1996 Young Microscopists competition. It was first published in Severs, N.J. (1996) Functional micro-anatomy of the cardiac muscle cell and its gap junctions. In *Recent Advances in Microscopy of Cells, Tissues and Organs* (P.M. Motta, ed.), University of Rome 'La Sapienza' (Proc. 2nd Int. Malpighi Symposium), and is reproduced here by permission.

Figure 5.3. Confocal projection image of an epidermal cell in a leaf of *Nicotiana benthamiana*, imaged after excitation at 488 nm, showing the distribution in the tubular cortical endoplasmic reticulum of the green fluorescent protein (GFP), which has been expressed in the plant from the potato virus X episomal vector, and targeted to the ER by a signal peptide sequence derived from patatin fused to its N terminus and the ER retention signal KDEL attached to its C terminus. The large round organelles are chloroplasts, visible by their autofluorescence, and the small brightly fluorescent bodies are thought to be Golgi bodies labelled with GFP. Image area ~ 58 × 83 µm. Reproduced by permission of Blackwell Science from Boevink, P., Santa Cruz, S., Hawes, C., Harris, N. and Oparka, K.J. (1996) Virus-mediated delivery of the green fluorescent protein to the endoplasmic reticulum of plant cells. *Plant J.* **10**: 935–941.

Figure 5.4. Composite confocal image of a whole young leaf of *Nicotiana benthamiana* which is systemically infected with potato virus X. The virus is expressing the gene for the green fluorescent protein (GFP) covalently attached to the viral coat protein, allowing the spread of the virus into areas adjacent to the secondary and tertiary veins to be imaged in intact leaves. The veins of the leaf have been separately labelled using a Texas Red dextran in the xylem. The leaf was dual-imaged after exciting at 488 nm for GFP (green) and at 568 nm for Texas Red (red). The entire leaf surface has been montaged from numerous smaller areas. Image courtesy of Alison Roberts (Scottish Crop Research Institute, Invergowrie, Tayside DD2 5DA, UK), who retains the copyright to this image.

Figure 5.5. Multichannel confocal projection image of part of a whole-mount zebrafish embryo stained with acridine orange, Evan's blue and eosin. The developing eye is evident in the top left region of the image. Sample prepared by Brad Amos (Medical Research Council Laboratory of Molecular Biology, Cambridge CB2 2QH, UK), and image courtesy of Duncan McMillan, Bio-Rad Microscience Ltd, who retains the copyright to this image.

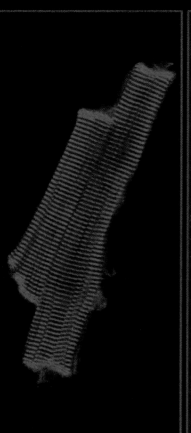

Figure 5.2

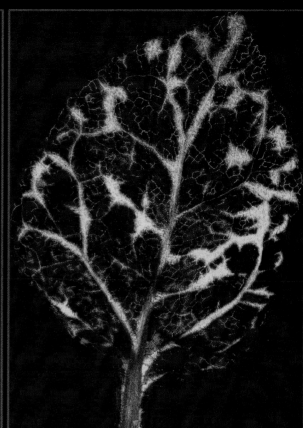

Figure 5.4

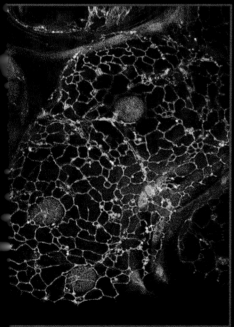

Figure 5.3

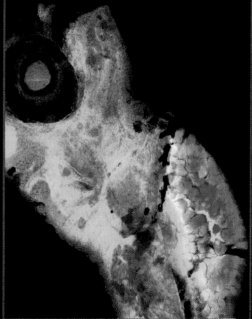

Figure 5.5

Colour Plate IV

Figure 5.6. Confocal multiparameter image of a *Drosophila embryo* at the blastoderm stage, triple-labelled for the transcription products of the genes *Hairy*, *Giant* and *Krüpple*, using immunoconjugates of lissamine rhodamine (red), fluorescein (green) and Cy-5 (imaged in the far red; displayed in blue), respectively. The three separate 8-bit confocal section images (R, G, and B) in the upper part of the figure are combined to give a 24-bit composite pseudocolour figure (RGB) below, which is also reproduced on the front cover of this handbook. The embryo is ~250 μm in length. First published in Paddock *et al.* (1993) Three-colour immunofluorescence imaging of *Drosophila* embryos by CLSM. *BioTechniques* **14**: 42–48, and in Bio-Rad Confocal Application Note 7 and reproduced here by permission of Eaton Publishing Co. and Bio-Rad Microscience Ltd, respectively, and courtesy of James Langeland who prepared the specimen and Stephen Paddock who created the images (both of Howard Hughes Medical Institute, University of Wisconsin, Madison, WI 53706, USA).

Figure 5.7. Confocal multiparameter image of an intact islet of Langerhans, triple-immunolabelled for insulin, somatostatin and glucagon using fluorescein, Cy-3 and Cy-5, respectively. First published in Brelje, T.C., Wessendorf, M.W. and Sorenson, R.L. (1993) Multicolour laser scanning confocal immunofluorescence microscopy: practical applications and limitations. *Methods Cell Biol.* Vol. 38, and in Bio-Rad Confocal Application Note 5, and reproduced here by permission of Academic Press and Bio-Rad Microscience Ltd, respectively, and courtesy of Todd Brelje (Department of Cell Biology and Neuroanatomy, University of Minnesota Medical School, Minneapolis, MN 55455, USA).

Figure 5.8. Confocal images taken at 5 sec intervals of a fertilization-induced calcium wave in a *Pisaster ochraceus* starfish oocyte that was microinjected with Calcium Green-1 conjugated to 10 000 MW dextran (Molecular Probes # C-3713), excited at 488 nm and imaged using a fluorescein filter set. The calcium concentration is indicated by the rainbow pseudocolour scale on the right, blue indicating a low and red a high concentration of free cytoplasmic calcium ions. The oocyte is ~130 μm in diameter. Image first published in Stricker, S.A., Centonze, V.E. and Melendez, R.F. (1994) Calcium dynamics during starfish oocyte maturation and fertilization *Devel. Biol.* **166**: 34–58, subsequently republished on the cover of Haugland, R.P. (1996) *Handbook of Fluorescent Probes and Research Chemicals* (6th Edn, Molecular Probes Inc.), and reproduced here by permission of Academic Press Inc. and Molecular Probes Inc., respectively, and courtesy of Stephen A. Stricker (Department of Biology, University of New Mexico, Albuquerque, NM 87131, USA).

Figure 7.4. An integrated circuit imaged using confocal microscopy. The brightness of the image represents surface reflectivity (maximum brightness projection), while surface height is encoded in pseudocolour, with red representing the highest parts and violet the lowest, as shown in the representation of the colour lookup table below the image itself. Reproduced by permission of Pergamon Press from Sheppard, C.J.R. (1986) Scanning methods in optical microscopy. *Endeavour* **10**: 17–19.

Figure 8.4. A cuvette filled with a safranin solution is illuminated from the lens at the upper right with green light at 532 nm from a frequency-doubled Nd:YAG laser, exciting fluorescence throughout the illuminated double cone of this low NA objective. The solution is also being illuminated from the lens at the lower left with infrared light at 1047 nm from a pumped mode-locked neodymium–yttrium lanthanum fluoride laser, which is exciting only a tiny spot of fluorescence by two-photon absorption at the exact focal point of the laser. Photograph courtesy of Brad Amos (Medical Research Council Laboratory of Molecular Biology, Cambridge CB2 2QH, UK), who retains the copyright to this photograph.

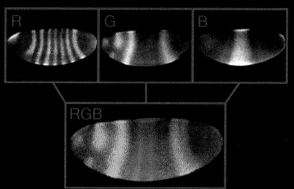

Figure 5.6

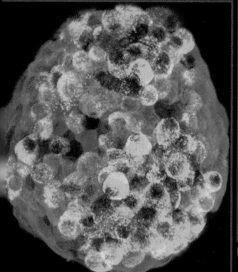

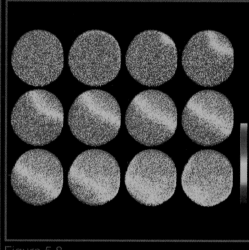

Figure 5.8

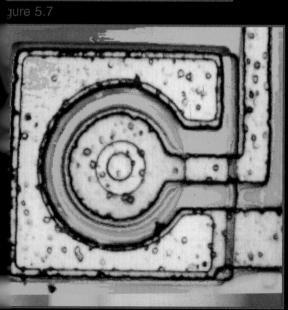

Figure 5.7

Figure 8.4

5 Confocal fluorescence microscopy

5.1 Instrumentation for multiparameter confocal fluorescence imaging

The modern confocal epi-fluorescence microscope differs little from a single-channel confocal reflection microscope, except that it is typically equipped with one or more lasers capable of delivering a suitable range of excitation wavelengths (*Table 1.1*), with between two and four photo-multipliers for confocal epi-fluorescence detection, which may be cooled to increase their signal-to-noise ratio, and with appropriate single and multiple dichroic beam splitters and emission filters. Optionally, there may also be a non-confocal transmission detector, thus creating a sophisticated multiparameter instrument capable of simultaneously or sequentially imaging a range of fluorochromes at two or more wavelengths. One great benefit of simultaneous multiple channel imaging using a confocal micro-scope is that the images obtained are in exact spatial register, since corresponding pixels in each image are measured synchronously during the illumination of a single voxel within the specimen. As mentioned in the previous chapter, this greatly facilitates any subsequent image processing that requires combining these images, such as multichannel pseudocolour display of multiply labelled specimens or fluorescence-ratio imaging of physiological parameters.

In choosing the configuration of a new confocal microscope, several fundamental questions have initially to be asked, and some hard choices have to be made. Is UV illumination required? What other lasers are needed to provide the necessary range of wavelengths for the planned investigations? How many photodetectors are required for any planned multiparameter imaging studies? Does the instrument need to be capable of video-rate imaging, and if so does the specimen emit enough light that a single-beam scanning instrument with a circular confocal illuminating aperture can be used, or does one have to employ the parallel data acquisition approach of slit-scan illumination? Does one require true confocal imaging with the highest possible axial resolution by using a circular confocal aperture, or does one need to compromise resolution in order to obtain fast scan rates with high light throughput by using a slit detector? Is a simple direct-view confocal system adequate, or is it important to have a system capable of digital image capture, and with software capable of subsequent sophisticated image processing? How user-friendly and robust is the software supplied with the system? Do the specimens to be studied require the microscope to be of upright or inverted configuration, or should the scanning system be interchangeable between both types?

Because the specifications of the commercially available confocal instruments change for the better with surprising rapidity, we refrain from making specific recommendations which would rapidly be outdated. Rather we would suggest the following broad principles. If the confocal microscope is to be used primarily for direct observation of blur-free fluorescence images, perhaps in place of a conventional full-field fluorescence microscope, or if it is to be employed primarily for imaging rapid dynamic events in fluorescently labelled living systems, then there would be wisdom in purchasing a relatively simple confocal microscope of the slit-scan or Nipkow disc type, with a video-rate output from a CCD camera. If, however, the confocal system is to form the core of a multi-user facility in which 3D multiparameter imaging is likely to be a significant requirement, then one should strive to obtain the most sophisticated, flexible and expandable single-beam confocal scanning system affordable within the available budget, giving appropriate thought as to who will manage and maintain the facility. Whatever the case, before making a choice, the purchaser is strongly recommended to undertake imaging, subsequent image processing, *and* quantitative sensitivity and bleach rate measurements on several alternative confocal microscope systems, using personal specimens under conditions that match as closely as possible those that will later be used experimentally.

5.2 Fluorescent preparation techniques techniques for biological specimens

Since a confocal fluorescence image contains information only from the focal plane, photons from the out-of-focus regions having been excluded by the confocal aperture, images of conventionally prepared specimens may appear less bright than when imaged using a full-field fluorescence microscope, although they will have an enhanced signal-to-noise ratio. It is thus often appropriate to label the specimen more intensely than one would for conventional fluorescence cytochemistry, by using higher concentrations of (immuno)cytochemical labelling reagents, or even by employing a triple-layer immunolabelling protocol (e.g. primary mouse monoclonal − secondary rabbit polyclonal anti-mouse IgG − tertiary fluorochrome-conjugated goat polyclonal anti-rabbit IgG) to achieve greater signal amplification.

To achieve high labelling efficiency and to avoid spurious artifactual cross-reactivities, it is important that the polyclonal antisera employed be affinity-purified, that the number of fluorochrome conjugates per immunoglobulin molecule be optimized, that appropriate blocking steps be used, and that suitable negative controls be examined. High fluorochrome labelling efficiency permits images to be collected at lower illuminating light intensities, thus avoiding unnecessary photobleaching. The golden rule of all confocal fluorescence microscopy is that one should use the lowest possible laser power that will still give a useful image. Increasing the laser intensity will, since the light is concentrated into a very small area at the focus, quickly saturate the fluorochrome (i.e. drive all the molecules into the excited state), with the consequence of increasing photobleaching throughout the entire depth of the specimen, while giving no increase in light output from the focal plane itself.

If the specimen is not to be mounted in a hardening mountant such as DPX (BDH Merck #36029) or Mowiol (Calbiochem #475904), the coverslip may be sealed into place with nail varnish or, preferably, with melted VALAP (a 1:1:1 mixture of vaseline, lanoline and hard paraffin wax, which forms a low melting temperature solid), applied with a cotton-tipped applicator or a fine brush. Since fluorochrome photobleaching is unavoidable, and is an even greater problem in confocal than in conventional microscopy, observation of fixed specimens should only be undertaken after mounting in a mountant containing a bleach retardant such as 100 µg ml^{-1} DABCO (1,4 diazobicyclo 2,2,2 octane; Sigma #D2522), 1 mg ml^{-1} N-propyl gallate (Sigma #P3130), 1 mg ml^{-1} p-phenylene diamine (Sigma #P1519), or a proprietary anti-fade reagent, Vectashield (Vector Laboratories #H-1000) being strongly recommended. A very good semi-permanent mountant for fluorescence microscopy can be obtained as follows: Make a solution of 20 g polyvinyl alcohol of average molecular weight 10 000, known commercially as Gelvatol (Monsanto Chemical Co.) or Airvol 205 (Air Products Ltd), in 80 ml of 140 mM sodium chloride, 10 mM sodium phosphate buffer, pH 7.2, by stirring for 16 h or heating in a boiling waterbath. Add 40 ml glycerol and stir for a further 16 h, then centrifuge at 12 000 to remove any undissolved solid. Remove the supernatant, check its pH is between 6 and 7, and store in 10 ml aliquots in the freezer. Add 1 mg DABCO to a 10 ml working solution of this preparation before use, and store in a syringe dispenser at 4°C. With living systems, these reagents cannot be employed, although ascorbic acid can be used at concentrations below 100 µg ml^{-1} as a free radical scavenger, provided it does not interfere with the physiology of the preparation. With living specimens, it is particularly important to minimize the excitation intensity and to avoid any unnecessary exposure of the specimen, both to avoid photobleaching and also because light with a wavelength of less that 500 nm is intensely cytotoxic. If a 3D data set is to be collected, the serial sections should be collected from front to back, so that the progressive fading, together with the inevitable depth attenuation, generates an appropriate depth cue by causing the rear of the object to appear fainter than the front. This effect, if quantified, may subsequently be corrected for mathematically.

Spatial resolution itself (discussed in Chapter 3) will be maximized by choosing a fluorochrome with a short excitation wavelength and a small Stokes' shift, although in practice this is usually a secondary consideration relative to the availability of suitable laser wavelengths and fluorochrome conjugates, and the need to avoid phototoxicity of living preparations. The range of fluorochromes and conjugates available for biomedical fluorescence imaging has expanded enormously over the last decade, and newcomers to the field will find the new Sixth Edition of the *Handbook of Fluorescent Probes and Research Chemicals* published by Molecular Probes (Haugland, 1996) an invaluable source of information. *Table 5.1* gives the absorption and emission maxima of many of the most commonly used fluorochromes, and more detailed information can be obtained from sources listed in the Further reading section.

5.3 Blur-free optical sectioning of single cells: applications in cell and molecular biology

Almost any fluorescently labelled specimen benefits from examination by CSOM, as evidenced by the large number of excellent papers employing this technique that have appeared in cell biological journals during the last decade. It is possible by confocal microscopy, for example, to visualize clearly microtubules in the vicinity of the nucleus of a stained cultured fibroblast (*Figure 5.1*, Colour Plate II), a region usually rendered uninterpretable by out-of-focus blur when imaged using a conventional full-field fluorescence microscope. Similarly, it is possible to distinguish with exceptional clarity the arrangement of structural proteins within an intact cardiac myocyte (*Figure 5.2*, Colour Plate III).

Early biological applications of vertical ($x\,z$) sectioning using CSOM included study of differences in endocytic rates between the apical and basolateral membranes of polarized Madin–Darby canine kidney (MDCK) epithelial cells, and measurement of the thickness of living cultured fibroblast cells for radiation dose–response experiments. For this latter study, the cells were made visible by the simple but effective and totally non-invasive strategy of confocal negative contrasting, by

Table 5.1. Absorption and emission maxima of some commonly used fluorochromes[a]

Fluorochrome name	Absorption maximum	Emission maximum	Principle application
Acridine orange	490	590, 640	Nuclear and RNA stain
Aequorin	–	469	Chemi-luminescent Ca^{2+} sensitive protein
Allophycocyanin	620	660	Immunoconjugates
AMCA (7-amino-4-methylcoumarin-3-acetic acid)	350	450	Immunoconjugates, DNA probes
7-Amino-actinomycin D	555	655	DNA stain
BCECF	500	530	pH-sensitive dye
BODIPY FL	503	512	Immunoconjugates, lipid probes
Calcein	495	520	Cell viability label
Carboxy-SNARF-1	518, 548 (acid)	587 (acid)/630 (base)	pH-sensitive emission ratio dye
Cascade Blue	376	423	Immunoconjugates
Chromomycin A3	450	570	DNA stain
Cy-3	556	574	Immunoconjugates
Cy-5	649	666	Immunoconjugates
DAPI (4',6-diamino-2-phenylindole)	372	456	Vital DNA stain
$DilC_{18}$	547	571	Long-term cell tracing
Ethidium bromide	545	610	DNA stain, dead cell assay
Fluorescein	496	518	Immunoconjugates
Fluo-3	506	526	Ca^{2+}-sensitive dye
Fura-2	335/362	510	Ca^{2+}-sensitive excitation ratio dye
Fura-red	488	660	Ca^{2+}-sensitive dye
Green fluorescent protein (GFP)	395 (475)	508	Gene expression, cell tracing
GFP bright mutant S65T	489	511	Gene expression, cell tracing
GFP blue mutant W7	433 (453)	475 (501)	Gene expression, cell tracing
Hoechst 33258 and 33342	345	478	Vital DNA stains
Indo-1	355	408/466	Ca^{2+}-sensitive emission ratio dye
Lissamine rhodamine B	570	590	Immunoconjugates
Lucifer Yellow CH	435	530	Neuronal tracer
C_6-NBD-ceramide	475	525	Vital Golgi stain
Phycoerythrin B	545	576	Immunoconjugates
Propidium iodide	530	615	DNA stain, dead cell assay
Tetramethyl rhodamine	554	576	Immunoconjugates
Texas Red	592	610	Immunoconjugates

[a] It should be noted that the excitation and emission spectra of fluorochromes are broad, and that the peak values will vary upon covalent linkage to antibodies and/or according to the chemical environment. The above figures are thus not absolutes, but for guidance only. Secondary maxima are shown in parentheses. Further details of these and other fluorochromes may be found in Mason (1993) and Haugland (1996) (see Further reading).

immersing the cells in culture medium containing a non-permeant high molecular weight fluorochrome-conjugated dextran, in which the cells appear as black voids within a fluorescent 'sea'.

CSOM has been widely used for fluorescence *in situ* hybridization (FISH) studies, using hapten-conjugated or directly fluorescent probes, at fluorescence intensities invisible by conventional optical microscopes. Its enthusiastic adoption for this purpose has been as much due to this sensitivity, and the ease of subsequent digital image processing of the data obtained, as for its out-of-focus blur rejection, since the specimens are usually thin. It can also be used to study gene expression at the cellular level, for example in cells transfected with constructs of the gene for the naturally fluorescent green fluorescent protein (GFP) from the jellyfish *Aequorea victoria* (*Figures 5.3* and *5.4*, Colour Plate III).

5.4 Applications in developmental biology, neurobiology and diagnostic cytopathology

Most studies using CSOM have employed high-magnification, high-numerical aperture objectives, but since these have a limited working distance, they are not suitable for looking deep inside thick specimens. However, much lower power objectives may be used to obtain good confocal fluorescence images over much larger areas, as exemplified in *Figure 5.4* (Colour Plate III), or from deep within bulkier specimens such as labelled vertebrate embryos, as in *Figure 5.5* (Colour Plate III). It is possible to image through half a millimetre or so of overlying fluorescently-labelled tissue, particularly when using far red or near infrared illumination which penetrates better than shorter wavelengths. Observations have been made non-invasively deep within vitally-stained living brain tissue and within the cornea of the living eye, and studies have been performed using video-rate confocal microscopy of the rapid physiological processing of fluorescent substances introduced into the blood through the glomeruli and tubules of a living kidney.

In diagnostic pathology, a particularly important application is that of making possible the rapid examination of biopsy specimens labelled with permeant fluorochromes, without the need to embed and section them, thus speeding diagnosis and subsequent treatment. In developmental biology, confocal microscopy permits one to follow the fates of individual microinjected or GFP-expressing cells, and to observe the reorganization of vitally-stained organelles and the expression of developmentally regulated genes during embryonic development.

5.5 Multiple wavelength imaging

Such studies of gene expression are particularly spectacular when using the confocal microscope as a multiparameter instrument for the simultaneous imaging of the location of several gene products, as exemplified by the cover photograph of this handbook and *Figure 5.6* (Colour Plate IV), where fluorescein, lissamine rhodamine and Cy-5, immunolabelling three distinct segmentally expressed *Drosophila* gene products, were sequentially excited by a multi-line argon–krypton ion laser using suitable filters. Another example of triple label imaging, this time of relevance to physiology, is an islet of Langerhans from a mammalian pancreas triple-stained for the three hormones insulin, somatostatin and glucagon using FITC, Cy-3 and Cy-5 immunoconjugates, showing clearly the segregation of expression of the three hormones between different cells of the islet (*Figure 5.7,* Colour Plate IV).

The use of high-quality interference filters is necessary to ensure maximal separation of the fluorochrome signals, those from Omega Optical and Chroma Technology being particularly recommended. However, since the emission spectra of many of the commonly used fluorochromes have long 'tails', this separation is never perfect. To a first approximation, such cross-talk may be corrected for mathematically. For this, one first determines the degree of bleedthrough of one fluorochrome, excited in isolation, into the other fluorescence channel(s), as a percentage of the signal recorded by its own channel PMT. After simultaneous multichannel imaging, that proportion of the first image is then subtracted from the image(s) recorded in the other channel(s). The separate corrected images may then be combined using suitable pseudocolours for display, as shown in the examples above. A sophisticated electronic method, involving high-frequency intensity modulation of the different excitatory laser lines, has recently been developed which effectively eliminates all cross-talk between fluorescence channels, but this is not yet available on commercial instruments.

5.6 Fluorescence ratio imaging of physiological parameters

A major application of multiparameter imaging, of enormous importance for the 3D determination of physiological parameters within living cells, such as transient changes in intracellular pH, free Ca^{2+} ion concentration and membrane potential, is simultaneous fluorescence emission ratio imaging, employing a physiologically sensitive dye which is excited by a single wavelength and imaged at two further wavelengths, only one of which is sensitive to the physiological parameter under investigation. A growing variety of fluorochromes appropriate for this task have recently been developed, requiring either ultraviolet or visible light excitation, of which the well-known examples are Indo-1 to quantify intracellular free calcium concentrations, and the SNAFL and SNARF pH indicator series.

The optical sectioning capability of CSOM may even dispense with the strict requirement for a *ratio* dye, if the primary need for this is to correct for variations in specimen thickness, thus allowing single wavelength imaging with such probes as Fluo-3 or Calcium Green for calcium ions (*Figure 5.8*, Colour Plate IV). With such non-ratiometric fluorochromes, 'pseudo-ratio imaging' may be achieved by co-injection of two non-ratiometric calcium-sensitive dyes such as Fura-Red and Fluo-3, both excited at 488 nm. In the presence of calcium ions, the former decreases its fluorescence emission in the red, while the latter increases its fluorescence emission in the green. Alternatively, Fluo-3 can be used with a second unrelated fluorochrome, insensitive to the parameter being measured, such as a fluorescent dextran, when the ratio of the reporter fluorochrome against the invariant fluorochrome will provide a large degree of correction for problems of fluorochrome exclusion or sequestration within different regions of the cell, provided that the partitioning and the bleaching rates of the reporter and invariant fluorochromes is similar. An improved pseudo-ratio imaging system has been proposed for calcium ions that avoids potential problems due to differential partitioning by using a covalent conjugate of Calcium Green-1 and Texas Red Dextran. Further details of these and other dyes suitable for confocal determination of physiological parameters are given by Mason (1993) and Haugland (1996) (see Further reading).

5.7 Combined confocal epi-fluorescence and non-confocal transmission imaging

As mentioned in Sections 2.3 and 5.1, the epi-fluorescence CSOM may optionally be equipped with a non-confocal detector for the scanned *transmitted* light. If the host optical microscope has darkfield, phase contrast or Nomarski DIC optical components, one may use these to collect a non-confocal scanned transmitted brightfield, darkfield, phase contrast or Nomarski image at the same time as one or more confocal epi-fluorescence images of the same field of view. The distribution of the fluorescent labels may then be displayed in suitable pseudocolours adjacent to or superimposed upon the monochrome transmission image. One model of the current (third) generation of commercial confocal microscopes is capable of simultaneously collecting five images from the same specimen, one non-confocal transmitted light image and four confocal epi-fluorescence signals.

5.8 Time-resolved fluorescence

In fluorescence, a molecule in an excited state rapidly decays to the ground state by emission of a photon. The resultant decay in light intensity follows an exponential relationship, where the fluorescence lifetime τ_f is the time taken for the intensity to drop from an initial value i to a value of i/e. For most fluorochromes, the fluorescence lifetime is of the order of a few nanoseconds, and is termed prompt fluorescence; if it is longer than this it is termed delayed fluorescence, and if very slow, phosphorescence. It is possible to generate confocal images in which the image displays the variations in fluorescence lifetime, rather than the intensity of the fluorescence signal. This approach has a number of useful properties and advantages, resulting from the fact that the fluorescence lifetime is specific to the probe, and is independent of illumination intensity, thus eliminating inaccuracies from photobleaching, shading and absorption. Firstly, time-resolved fluorescence imaging can give chemical information on transient effects such as the lifetime of excited states, and the capture and emission cross-sections, which can be used in studies of, for example, molecular dynamics and energy transfer. Secondly, time-resolved fluorescence imaging can be used in multi-labelling experiments to distinguish different probes that fluoresce at the same wavelength but with different fluorescence lifetimes. This approach can be used instead of, or together with, spectral filtering. Fluorescence lifetime can be sensitive to ionic concentration, so that it can also be used as a measure of pH, or of Ca^{2+} or Na^+ concentration. Further, fluorescence lifetime is directly proportional to the fluorescence quantum efficiency, so that it can be used to correct for changes in quantum efficiency in quantitative fluorescence measurements.

Time-resolved fluorescence can be performed in either the time or the frequency domain. In the time domain, a pulsed laser is used and the fluorescence intensity measured as a function of time. This is most often achieved in confocal microscopy by time-gated detection, in which two (or more) detectors are used to detect the fluorescence signal within particular time windows. Assuming an exponential fluorescence decay, the fluorescence lifetime can then simply be calculated from the ratio between the two signals. Another technique is time-correlated single-photon counting, in which the time decay curve is directly measured. However this approach suffers from the drawback that it is rather slow.

In the frequency domain, the laser is modulated and the phase of the fluorescence signal measured using phase-sensitive detection.

5.9 Conclusion

The range of applications of confocal fluorescence microscopy in the biomedical sciences is now enormously diverse, as can be seen by scanning the pages of any of the leading journals of cell and developmental biology. The examples given in this chapter are not intended to be comprehensive, but rather were chosen to be visually clear, conceptually straightforward, and interpretable without requiring specialist knowledge of the biological systems under investigation.

6 Biological applications of confocal reflection microscopy

6.1 Colloidal gold imaging

Colloidal gold is now the immunocytochemical labelling reagent of choice for subsequent electron microscopic examination. Such gold probes of 20–40 nm diameter may also be clearly seen both by confocal reflection SOM and by video enhanced contrast microscopy. Since the reflection signal obtained from 40 nm particles is strong and resistant to photobleaching, immunogold labelling is sometimes preferred over immunofluorescence labelling for 3D image collection from fixed specimens, where photobleaching of fluorochrome molecules during serial optical sectioning might otherwise present serious problems. The slightly higher spatial resolution observed in practice from confocal reflection as compared with confocal fluorescence imaging also suggests that immunogold labelling may be the method of choice for 2D imaging.

Since silver has a higher reflectivity than gold, even better staining (although not resolution of individual particles) may be obtained by silver enhancement after saturation immunolabelling with small 1 or 5 nm gold conjugates, which themselves would not provide a significant reflection signal.

While individual 20 or 40 nm gold particles on a cell surface can be seen by confocal reflection microscopy after the cell has been fixed, cleared and mounted in a mountant of suitably high refractive index such as glycerol ($n = 1.474$) or DPX (*Figure 6.1*), the situation is less straightforward when imaging gold conjugates on living cells immersed in a tissue culture medium of refractive index considerably lower than that of the cell. Here reflections from the cell itself, particularly from the medium–plasma membrane interface, are significant and make it difficult to distinguish individual gold particles. Such reflections provide a limit to the minimum size of particle which may be detected. It should

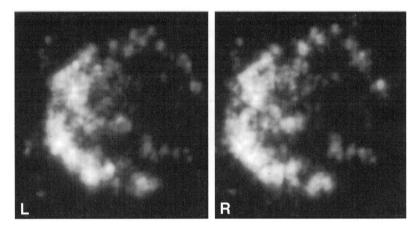

Figure 6.1. Stereoscopic image pair showing a 3D confocal reflection image of a rat thymocyte labelled with anti-CD43 and a 40 nm gold- conjugated secondary antibody, fixed after incubation at 37°C to induce partial capping of this surface antigen. The cell diameter is ~10 μm. Reproduced by permission of Elsevier Trends Journals from Shotton, D.M. and White, N.S. (1989) Confocal scanning microscopy: three-dimensional biological imaging. *Trends Biochem. Sci.* **14**: 435–439.

be noted that, for a particle which is small compared with the wavelength, the intensity of the light scattered is proportional to the fourth power of the diameter, so that while particles as small as 5 nm in diameter have been imaged, this has only been achieved under ideal circumstances with fixed preparations and subsequent image averaging to enhance the SNR.

In principle, it should be possible to see gold particles on living cells equally effectively by confocal transmission brightfield microscopy, as can be achieved by brightfield video-enhanced contrast transmission microscopy ('nanovid' microscopy), since they are completely opaque while the surrounding cell is transparent. However, the alignment problems inherent to confocal transmission microscopes has prevented their commercial development, and thus has severely limited the number of laboratories able to explore such applications.

6.2 Confocal reflection imaging of surface replicas and gold-coated SEM specimens

Surface replication using platinum–carbon or tantalum–tungsten is a widely used technique for high resolution transmission electron microscopic (TEM) imaging of biological tissues and macromolecules after freeze fracture or deep etching. Such surface replicas are also excellent specimens for confocal reflection imaging, enabling the 3D topographical detail of the surface to be determined with greater ease than by parallax calculations from stereoscopic pairs of transmission electron micrographs obtained by specimen tilting.

This opens the way for sophisticated correlative microscopy, in which, for example, a cell is first imaged by confocal reflection or fluorescence microscopy in the living state after immuno-labelling of its surface antigens, and is then rapidly frozen, freeze-fractured and replicated. This replica, after clearing free the biological material and mounting on a grid, may then be studied by high-resolution TEM, which, using the label-fracture technique, will also reveal the positions of the surface immunogold particles. It may then be immersed in oil and imaged by confocal reflection microscopy to determine its 3D topography. With suitable image processing, these different image data may be brought together. By the same token, SEM specimens prepared by gold sputter coating provide excellent confocal reflection images, albeit with lower spatial resolution, with similar possibilities for correlative microscopy.

6.3 Interference reflection contrast imaging

Confocal SOM is also capable of generating reflection interference contrast images that may be used, for example, to visualize the focal adhesions made by living cells on a glass coverslip. In principle, the confocal microscope should be superior to the conventional optical microscope for interference reflection microscopy, because of the confocal rejection of spurious reflections from the upper cell surface. However, the published examples of such confocal applications are as yet too few to evaluate.

7 Industrial applications of confocal microscopy

7.1 Micrometrology

Scanning optical microscopy is particularly suitable for making quantitative dimensional measurements, as a consequence of the electronic form of its image signal. This ability is enhanced for the confocal instrument because of the improved axial resolution capabilities. A further advantage of the confocal microscope for metrology is that the plane of measurement can be accurately located. There is a wide range of applications for such methods, including measurement of fibres, small machined parts, magnetic recording heads, and semiconductors.

For accurate measurements, the magnification of the confocal images needs to be calibrated, as the nominal magnification of most objective lenses is not sufficiently exact. This can be achieved by using a calibration specimen, or a translation stage with a laser interferometer. Confocal microscopes with a linear CCD detector (Section 1.5) are also suitable for making dimensional measurements, as the scale is then determined by the geometry of the CCD array.

In order to measure the size of a feature, its edges must be located. This can be achieved with a sensitivity much greater than the optical resolution, perhaps to within a few nanometres. For absolute measurement, the position of the edge relative to the intensity profile across its image must be determined. For a light–dark boundary imaged using a fully incoherent imaging system, the edge is located at the position where the intensity is midway between the light and the dark regions. This is called the edge-setting condition. The problem is that in practical non-confocal systems, imaging is partially coherent, so that the relative intensity at the edge is no longer exactly one half, and, even more importantly, it varies according to the presence of aberrations or defocus. However, in confocal laser scanning microscopes, where reflection imaging is fully coherent, the relative intensity at the edge is one quarter of that in the bright region (since from symmetry considerations the rela-

75

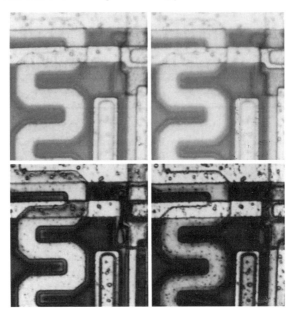

Figure 7.1. An integrated circuit imaged in reflection. The top images are conventional images, and the lower ones confocal. The improved contrast of the confocal images is apparent. The left-hand images were focused on the sinusoidal structure in the left half of the image, while the right-hand images were focused on the horizontal bar about one third of the way down the image. This illustrates how focus position is easily determined in confocal microscopy. Reproduced by permission of Pergamon Press from Sheppard, C.J.R. (1986) Scanning methods in optical microscopy. *Endeavour* **10**: 17–19.

tive amplitude at the edge is one half and the intensity is given by the square of the amplitude) and is independent of the presence of aberrations. Thus the edge can be located absolutely. It should be noted that the visual position of the edge, which occurs at a relative intensity of one half, is thus spatially separated from the true position, so that visual inspection of confocal images yields an under-estimation of size for a bright object on a dark field, and an over-estimation of size for a dark object on a bright field.

A further advantage of confocal microscopy is that as a result of the depth discrimination property, the level at which the microscope is focused is immediately apparent. This is illustrated in *Figure 7.1*, which shows images of an integrated circuit obtained by conventional and confocal reflection microscopy.

7.2 Surface profiling and surface examination

If a planar surface is scanned axially through the focus of a reflection confocal microscope, the intensity reaches a maximum when the specimen surface coincides with the microscope focus (see *Figure 2.2*). This is true even for a sloping surface. Thus for a specimen which is not planar, the local surface height can be determined by locating the position of maximum intensity during axial scanning (see *Figure 2.3*). Dedicated profiling instruments can perform this axial scanning, recording the position and perhaps also the value of the maximum intensity, which is a measure of the surface reflectivity. However, in general purpose confocal microscopes, it is necessary first to record a complete confocal 3D data set, and then to extract the profile information subsequently, unless the geometry of the specimen is such that a single $x z$ image is sufficient to provide the required data. As with the generation of 3D projections, a range of alternative algorithms are possible, the most straightforward being the peak detection algorithm. Reasonable results are obtained simply by selecting the maximum voxel value, but in this case sensitivity is limited to the axial scan sampling period. Much better sensitivity may be achieved by using a parabolic fit to the individual voxel intensities (*Figure 7.2*). As with transverse measurements, the axial sensitivity of surface profiling can be much greater than the axial resolution, and in practice the surface position can be determined to within the order of 1 nm. However, peak detection is inherently noisy, so that other algorithms may be preferable. One alternative is the determination of the centre of gravity of the axial intensity variation, but this suffers from noise originating from the highly weighted regions far from the surface. For this reason, a better algorithm is the determination of the centre of gravity of some power of the axial intensity distribution, the square being a compromise which is found to produce good surface profiles. Surface profiling sensitivity can be further improved using confocal interference methods, but at present there are no commercial instruments based on this approach.

An example of a surface profile is shown in *Figure 7.3*. The profile is shown as an isometric view, the brightness of the reconstruction representing reflectivity. This profile was generated from a 3D confocal data set by parabolic fitting. Another example of a profile is shown in *Figure 7.4* (Colour Plate IV). In this case the height was determined simply from the position of the brightest sample in the axial scan. The height

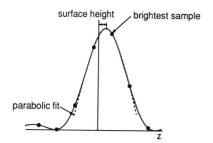

Figure 7.2. The axial intensity variation in the confocal image through a point on a surface. A series of samples of the intensity is shown. The position of the brightest sample gives an approximate measure of the surface height. A much more sensitive measurement is given by fitting a parabolic curve through the three brightest samples.

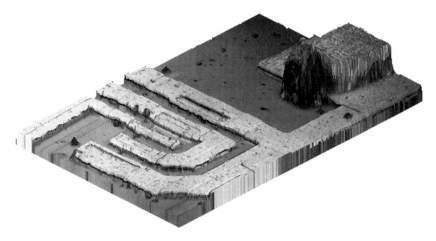

Figure 7.3. A 3D reconstruction of a surface profile of an integrated circuit. The surface is shown as an isometric view, the brightness of the image representing the surface reflectivity. The area of the region is 400 × 250 μm. The silicon surface shows up as midgrey, with bright lines of metallization, which are about 1.5 μm thick. The bonding pad in the upper right-hand corner is about 10 μm thick. The reconstruction was performed on a 3D data set (courtesy of G. Cox) using a parabolic fit, using software developed by I. Huxley and C.J.R. Sheppard.

and maximum brightness projection images were combined in a single pseudocolour image, in which the colour represents surface height, in the same way as in a geographical map.

7.3 Thin film profiling

As a generalization of surface profiling, confocal microscopy can be used to investigate multilayer structures. The true surface of a substrate can be observed through a surface coating, and the thickness of a thin film coating a substrate can be determined by observing the two peaks in the axial intensity variation. For such studies, it is necessary to know the refractive index of the film in order to take into account the refractive depth distortion of the image, as discussed in Chapter 3. The resolution of this method, which can be extended to the observation of more complicated axial structures, is limited to about a wavelength.

Figure 7.5 shows theoretical axial images through a sample consisting of a thin film of silicon dioxide on a silicon substrate. A peak in signal is recorded for reflection from both the top surface of the thin film and the film–substrate interface. The images were calculated using a rigorous high-aperture theory. It should be noted that as confocal reflection imaging is a coherent process, the relative phase of the reflections is important in determining the detail in the images. *Figure 7.6* shows

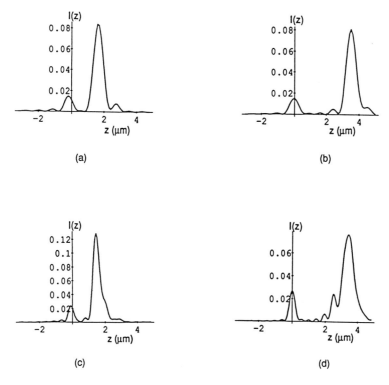

Figure 7.5. Theoretical image of a thin film structure of silicon dioxide on a silicon substrate, calculated using a vigorous high-aperture theory: (a) dry objective, NA = 0.80, film thickness d = 2.5 µm; (b) dry objective, NA = 0.80, film thickness d = 5.5 µm; (c) dry objective, NA = 0.95, film thickness d = 2.5 µm; and (d) dry objective, NA = 0.95, film thickness d = 5.5 µm. Reproduced by permission of the Optical Society of America from Sheppard, C.J.R., Connolly, T.J., Lee, J. and Cogswell, C.J. Confocal imaging of a stratified medium. (1994) *Appl. Opt.* **33**: 631–640.

experimentally recorded axial images of a thin film, which exhibit two pronounced peaks representing the two interfaces, and agree quantitatively with the theoretical predictions.

7.4 Microscopy of semiconductor materials and devices

One of the major industrial application areas for confocal microscopy is in the investigation of semiconductor devices. Optical methods have many advantages over electron microscopy for semiconductor studies, including their non-invasive nature, which avoids the generation of surface defects and contamination, and the ability to image through non-conducting passivation layers. These applications include dimensional metrology of device structures, including so-called critical dimension

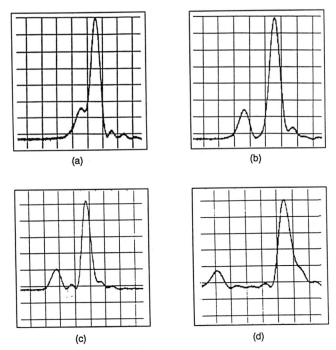

Figure 7.6. Experimental axial images, recorded using a 0.95 NA objective, of silicon dioxide films on a silicon substrate. The film thicknesses are (a) 1.0 µm; (b) 2.5 µm; (c) 3.0 µm; and (d) 5.5 µm. Horizontal scale: (a), (b) and (d), 1 division represents 0.81 µm; (c) 1 division represents 0.89 µm. Reproduced by permission of The Optical Society of America from Sheppard, C.J.R., Connolly, T.J., Lee, J. and Cogswell, C.J. Confocal imaging of a stratified medium. (1994) *Appl. Opt.* **33**: 631–640.

analysis. The resultant images can be compared either with a reference image, or with computer design data. Specialized commercial instruments have been developed for such applications.

The helium–neon laser is often used for microscopy of semiconductors as the red light can penetrate below the surface into the bulk, and infrared illumination is also useful, as it can be used to image wafers in transmission. Polarization microscopy can be used to investigate crystal imperfections, and the adhesion of metallic contacts observed through the wafer.

The confocal microscope can also be operated in the photoluminescent mode, analogous to fluorescence microscopy of biological specimens. Luminescence can give information about spatial variations in excitation states, binding energies, band structures, molecular configurations, structural defects, and the concentration of atomic and molecular species. Commercial confocal Raman imaging systems are also available.

In the OBIC (optical-beam-induced current) method, the focused laser beam is used to generate electrical carriers in the semiconductor and the resultant current is monitored to produce an image. The strength of the current in an OBIC image varies as a function of diffusion length, junction depth, surface recombination velocity, or the presence of crystal

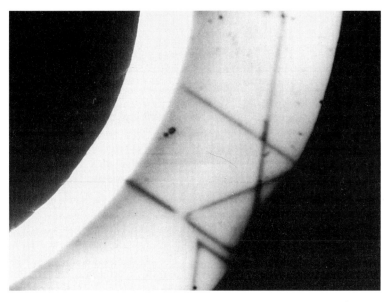

Figure 7.7. A silicon transistor imaged using the OBIC mode, with contrast enhancement. The dark lines represent defects, where the photocurrent generated by the laser illumination is reduced by recombination. Reproduced by permission of SPIE, The International Society for Optical Engineering, from Sheppard, C.J.R. (1982) Applications of scanning optical microscopy. *Proc. SPIE* **368**: 88–95.

defects such as dislocations, grain boundaries or trapping centres. Applications have included the investigation of photoconductors, photodiodes, LEDs, laser diodes, photocathodes and solar cells, discrete electronic devices and integrated circuits. *Figure 7.7* shows an OBIC image of a transistor. The dark lines are defects in the crystal structure, where the photocurrent is reduced by recombination. The OBIC technique has also been used to observe the formation of microplasmas and leakage channels. Logic states and the gain of integrated circuit transistors can be determined, and the transistors can be switched by the laser to test logic functions. Monitoring of the effect of altering the supply voltage upon laser illumination can be used to locate latch-up sites, regions in a complementary metal-oxide-semiconductor (CMOS) device where there is an unwanted parasitic electronic interaction.

The OBIC method can be performed on plain semiconductor materials either by preparing a Schottky barrier junction or a pressure or electrolytic contact. Alternatively, using a modulated beam, the surface photovoltage can be detected capacitively, without metallization or physical junction formation.

8 The future of confocal microscopy

8.1 Super-resolution

As described in Chapter 3, confocal microscopy can result in resolution greater than that achievable using conventional optical imaging (super-resolution). Even greater super-resolution advantages than those achievable by normal confocal imaging may be obtained by combining confocal and digital deconvolution methods. Generally, in order for digital processing methods to result in an increase in the spatial frequency bandwidth of the image, it is necessary for some extra information to be provided which is used to constrain the image deconvolution. This may be in the form of *a priori* information such as a known PSF, non-negativity of the object intensity, or the finite size of the object. Alternatively some other parameter, such as signal-to-noise ratio, can be traded off in order to improve resolution.

A different method for improving resolution, demonstrated in the laboratory but yet to be developed into a practical instrument, employs imaging with an array of detectors which record both the on-axis and the off-axis light imaged from the illuminated voxel. In theory, by using new algorithms to process the additional information obtained, an increase in lateral resolution is obtainable. For coherent imaging, this increase (defined in terms of the FWHM of the image of a point object) is twice the resolution of a conventional microscope and 1.4 times that of confocal reflection imaging, and for incoherent (fluorescence) imaging, a resolution increase up to four times that of a conventional microscope, and 2.8 times that of confocal fluorescence, is obtainable in principle.

Resolution can also be increased by the use of optical filters, which modify the transmission properties of the objective lens, thus sharpening the point spread function. Again, this method has been demonstrated in principle, but is not offered on commercial systems.

A much more direct way of increasing confocal resolution which has recently been demonstrated, with exciting possibilities for future practical implementation on commercial confocal microscopes, is that of directly 'engineering' the shape of the fluorescence emission point spread function. This relies upon the principle of *depletion of fluorescence by stimulated fluorescence emission*, in which the regions of the specimen immediately adjacent to the peak position of the fluorescence excitation beam, overlapping the edges of the illuminating Airy disc, are irradiated with light of a wavelength equal to that of the peak fluorescence emission of that fluorochrome. By suppressing fluorescence emissions in this annular region, and thus limiting the dimensions of the point from which fluorescence is actually emitted to the central region of the illuminating Airy disc, this point spread function 'engineering' can result in resolution enhancements of four- or five-fold, giving attainable lateral resolutions of 30–50 nm!

8.2 Near-field scanning optical microscopy

Discussion of resolution enhancement in scanning optical microscopy would not be complete without mention of a quite different method, near-field scanning optical microscopy (NSOM), a scanning probe microscopy technique closely akin to scanning tunnelling microscopy (STM), that in principle can give spatial resolution of better than one tenth of a wavelength. All the types of SOM discussed so far in this handbook are 'far-field' imaging techniques, limited by the constraints of Fraunhofer diffraction. By contrast, NSOM produces images by illuminating the specimen with light emerging from (or collected by) a minute aperture in the tip of a highly tapered probe positioned in the 'near-field' region, at a distance of less than a wavelength from the specimen surface. At this proximity, the radiation emitted from the sub-wavelength diameter aperture has dimensions similar to the size of the aperture itself, and independent of wavelength. If the transmitted or fluorescently emitted light is then imaged on to a detector using conventional far-field optics, an image can be produced with enhanced spatial resolution. However, for this to occur, the distance of the probe from the specimen surface must be kept constant, for example at 20 ± 1 nm for a 50 nm diameter aperture used with 500 nm light, since the signal intensity varies exponentially with distance, as in STM. This is achieved by feedback control either of the tunnelling current, or the capacitance or shear force between the probe and the surface. Scanning the probe thus simultaneously generates both optical information (absorbence, fluorescence) and 3D topographical information about the specimen surface, to a higher resolution than can be obtained by far-field CSOM. Its applicability to wet biological specimens has yet to be demonstrated, however, and its biological applications, as with STM, are more likely to be in the macromolecular than the cellular domain.

8.3   4 pi and theta confocal microscopy

While the foregoing discussion has concentrated on methods of enhancing lateral spatial resolution beyond that of normal confocal imaging, a more significant advance would be to improve the confocal axial resolution, which is usually some three or four times poorer that the lateral resolution (*Figures 3.8b* and *8.1a*). It is to this end that the techniques of 4 pi confocal microscopy and theta confocal microscopy have been developed.

In a confocal microscope, both transverse and axial resolutions are limited by the aperture of the objective lens. It is thus intriguing to consider theoretically the effects of increasing the aperture until the specimen is illuminated by a complete converging sphere, when obviously the resolution is isotropic. The resolution in this limiting case is improved by factors of about 1.3 in the transverse direction and about 3.4 in the axial direction, compared with normal confocal imaging using a high aperture objective. In practice, of course, it is impossible for geometrical reasons to achieve this, but broadly any increase in the coverage of the complete sphere is advantageous.

To this end, the 4 pi method uses two aligned opposed objectives to increase angular coverage. A result of the increased but yet incomplete coverage is a desirable tightening of the 3D PSF along the optical axis, but with the appearance of strong sidelobes in the axial response. However, these can be removed by subsequent image processing. There is a great technical problem with this technique in aligning the two objectives so that their two foci coincide exactly. It is necessary that the phases of these two waves also match. Just as in normal confocal transmission microscopy, it is very difficult to achieve this when focusing through a real biological object, since refractive index changes distort the phase boundaries, and this remains a major limitation of the method. One way of reducing this problem is to use the double pass method, as described in Section 2.3. In a 4 pi microscope, the double objective method can be used either for illumination or collection, or for both. However, in practice it is very difficult to achieve satisfactory alignment for both together, and usually the 4 pi method is used only on the illumination side.

In the confocal theta microscopy method, two objectives are again used, but here the axis of the collection optics is offset relative to the illumination system by an angle θ, typically close to 90° (*Figure 8.1b*). It is similar to arrangements which have been widely used in flow cytometry, Raman spectroscopy and luminescence microscopy. Since the PSFs of the illumination and detection objectives are inclined relative to one another (*Figure 8.1c*), the axial resolution of the illumination is improved by the transverse resolution of the detection optics (*Figure 8.1d*). This method does not rely on phase coherence of the two paths, so

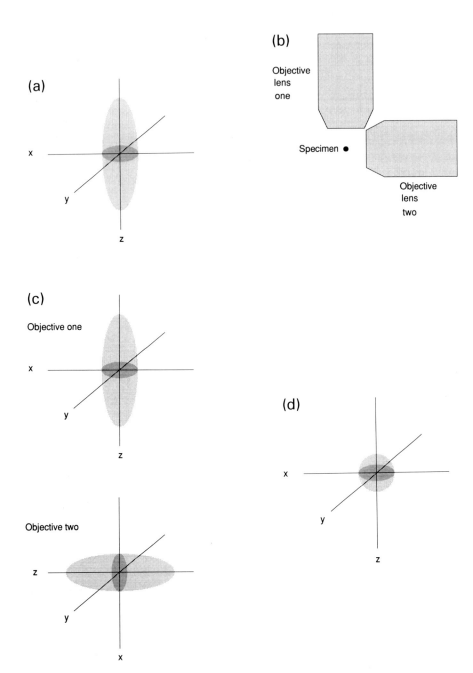

it is much easier to implement in practice than the 4 pi approach. In the prototype, water-immersion objectives were used to minimize refractive effects, and, because of purely geometric considerations of positioning two objectives confocally when oriented orthogonally, long working dis-

Figure 8.1. The principle of confocal theta microscopy. (a) Diagrammatic representation of the resolution volume of a conventional single objective epi-illumination confocal microscope (equivalent to the limits of the 3D point spread function, *Figure 3.8b*, but rotated by 90° so that the optical axis is vertical). The axial resolution is about 700 nm, while the in-plane resolution is about 200 nm. (b) The geometric arrangement of the two objective lenses employed in theta confocal microscopy. Because of physical constraints, the matched objectives used in the prototype theta confocal microscope were medium-power water-immersion objectives with a long working distance, and the angle θ between them was slightly larger than the 90° shown here. (c) The mutually inclined 3D PSFs of the two objectives used in theta confocal microscopy. (d) The resulting PSF of the theta microscope, obtained by multiplying the two PSFs shown in (c). Resolution is now approximately equivalent along all three axes. Reproduced by permission of Springer-Verlag from Shotton, D.M. (1995) Electronic light microscopy – present capabilities and future prospects. *Histochem. Cell Biol.* **104**: 97–137.

tance objectives were used. Hence the magnification and aperture of the objectives was limited, so that in practice there was no overall improvement in the resolution volume compared with that of a single high-aperture objective confocal system. However, the resulting PSF was almost isotropic (*Figure 8.1d*). This, coupled with the long working distance, makes this form of theta microscopy attractive for some applications such as observation of intact embryos. More recent developments of this technique, using mirrors to avoid geometric constraints, have permitted the use of high NA objectives.

8.4 Two-photon excitation and other non-linear confocal techniques

In fluorescence microscopy, the electronic state of a dye molecule is excited by absorption of a single photon of light at one wavelength, and decays back to the ground state with the emission of a single photon at a slightly longer wavelength; the difference between these wavelengths is defined as the Stokes' shift (*Figure 8.2a*). One of the major limitations

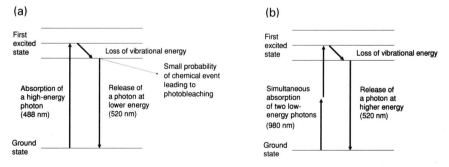

Figure 8.2. (a) Simplified diagram of the electron events occurring during conventional fluorescence excitation of fluorescein. (b) The principle of two-photon excitation. Reproduced by permission of Springer-Verlag from Shotton, D.M. (1995) Electronic light microscopy – present capabilities and future prospects. *Histochem. Cell Biol.* **104**: 97–137.

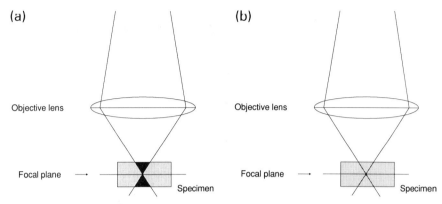

Figure 8.3. Diagram to illustrate the advantage of two-photon excitation microscopy with respect to confocality and specimen photobleaching. (a) The illuminated cones above and below the specimen during normal single-beam confocal excitation of a single point in a 3D specimen. To a first approximation, ignoring absorption effects, the total power through the specimen at any height is constant, causing the same amount of fluorescence excitation and bleaching at all levels. The out-of-focus fluorescence emissions are excluded from the final image by the confocal aperture. (b) In two-photon excitation, the photon flux is sufficiently intense to excite fluorescence emissions only at the focal point itself. Since no emissions occur elsewhere, the image acquired is intrinsically free of out-of-focus blur, and a confocal aperture is unnecessary. Photobleaching is likewise limited to the focal spot. Reproduced by permission of Springer-Verlag from Shotton, D.M. (1995) Electronic light microscopy – present capabilities and future prospects. *Histochem. Cell Biol.* **104**: 97–137.

of fluorescence microscopy, whether applied conventionally or confocally, is the progressive bleaching of the fluorochrome during prolonged exposure. Unfortunately, when focusing at a particular depth within a transparent fluorescently labelled 3D specimen, fluorochrome molecules throughout the whole of its thickness are excited (*Figures 8.3a* and *8.4*, Colour Plate IV). Although the illumination is weaker in out-of-focus planes, the duration of exposure during confocal scanning is longer, so that the integrated intensity of illumination at the fluorescence excitation wavelength remains constant. This results in almost equal photobleaching throughout the entire depth of the specimen upon every scan (*Figures 8.3a* and *8.5d*), and thus limits the number of confocal optical sections which can be recorded.

In two-photon fluorescence microscopy, the dye molecules are excited by the simultaneous absorption of two photons of longer wavelength, each of which alone has insufficient energy to cause a transition to the excited state, the strength of the two-photon absorption being proportional to the square of the illumination intensity (*Figure 8.2b*). This has a number of important consequences. Firstly, the intensity of illumination can be made sufficiently high to excite two-photon fluorescence only in the focal region, so that fluorescence emission imaging is identical to that of a confocal microscope, even without the use of a confocal aperture (*Figures 8.3b* and *8.4*, Colour plate IV). Thus optical sectioning is obtained, coupled with a strong detection efficiency. The optical performance of the system can in principle be improved by using a confocal

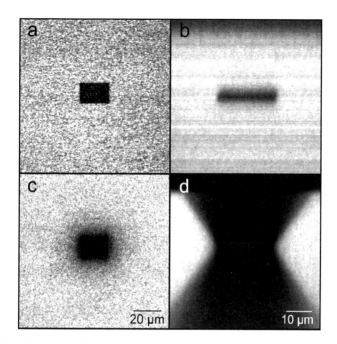

Figure 8.5. Photobleaching several microns below the surface of a three-dimensional gel of FITC–dextran induced by two-photon absorption of light at 760 nm from a pulsed titanium–sapphire laser (a and b) or by single photon absorption at 488 nm (c and d). (a and c): $x\,y$ sections; (b and d): $x\,z$ sections. In the case of two-photon excitation, photobleaching is limited to the illuminated area in the focal plane, while with normal single-photon excitation the photobleaching extends in a double cone throughout the specimen, as shown diagrammatically in *Figure 8.3a*. Reproduced by permission of the Royal Microscopical Society from Kubitscheck, U., Tschödrich-Rotter, M., Wedekind P. and Peters R. (1996) Two-photon scanning microphotolysis for three-dimensional data storage and biological transport measurements. *J. Microsc.* **182**: 225–233.

aperture, but this suffers from the disadvantage of reducing the detection efficiency. Secondly, since two-photon excitation only occurs at the focus, bleaching at points situated away from the focus is negligible (*Figures 8.3b* and *8.5b*). Two-photon fluorescence microscopy thus gives optical sectioning equivalent to that of a confocal microscope, but with much reduced bleaching. Furthermore, since the red or near infrared illuminating light used for two photon excitation has approximately twice the wavelength of that employed for single photon excitation, depth penetration into biological or materials specimens is greatly enhanced, permitting imaging from deeper within the specimen than could be achieved confocally. In addition, phototoxicity due to the long wavelength light is negligible in comparison with that caused by single photon excitation of the same emissions. On the other hand, the increased wavelength results in poorer spatial resolution.

Two-photon fluorescence is intrinsically a very weak process, because of its requirement for simultaneous absorption of two long-wavelength photons, thus requiring a very high illumination intensity to be effective. However, the allowable *average* illumination power that can be tolerated by the specimen is limited by specimen heat dissipation and single-photon fluorescence photodamage effects. Since the two-photon emission intensity is proportional to the square of the illumination intensity, the two-photon signal for a given *average* power can be greatly increased by using a pulsed laser beam, without causing photodamage. The two-photon signal, for a given average power in the absence of saturation, is proportional to the reciprocal of the duty cycle, t_p/T_p, where t_p is the pulse duration (typically 100 femtoseconds) and T_p is the repetition time. On further increasing the repetition time, for a given pulse duration and a constant average power, the pulse intensity will increase and the fluorochrome excitation at the focus will saturate. Thus the repetition time should not be made too long. However, if it is made shorter than the fluorescence lifetime, t_f, a second pulse excites the fluorochrome before it has time to fluoresce, thus reducing the attainable signal level. As a result, the optimum repetition time is approximately equal to the fluorescence lifetime, which in practice is a few nanoseconds.

Typical mode-locked lasers satisfy these requirements quite well. Pulse-lengths of about 100 fs are available. Much shorter pulses have been produced in the laboratory, but they are not suitable for microscopy without special techniques, as a result of pulse broadening due to dispersion in the optical elements. In practice, mode-locked dye lasers or, preferably, titanium–sapphire lasers are suitable. The titanium–sapphire laser can be tuned over the wavelength range from 700 to 1050 nm, which corresponds to excited state energies in the single-photon range from 350 to 525 nm, matching the absorption maxima of many commonly-used fluorochromes, including the Ca^{2+} sensitive dyes Indo-1 and Fura-2 which normally require UV excitation for single-photon fluorescence (*Table 5.1*). Unfortunately, despite growing demand, these lasers are still very expensive at present, but hopefully cheaper systems based on semiconductor diode lasers will soon become available. At the time of writing, the first commercial two-photon imaging system has just been developed.

There is a wide range of other non-linear optical techniques which can be used in laser scanning microscopy. Those which have been reported include second harmonic generation, two-photon absorption, and coherent anti-Stokes Raman scattering. In second harmonic generation, two photons combine to produce a single high energy photon, which is then detected. In two-photon absorption microscopy, molecules are excited, and the population of the excited state affects transmission of a probe beam. These methods can give information concerning crystal orientation and perfection, molecular structure and surface properties. In general, these non-linear methods also result in optical sectioning without use of a confocal pinhole.

8.5 Conclusion

Confocal scanning optical microscopy is undoubtedly the most significant advance in optical microscopy within the last two decades, and has become a powerful tool for the cellular and developmental biologist, the materials scientist and the microelectronics engineer. It is entirely compatible with the range of conventional light microscopic techniques, and, at least in scanned-beam instruments, may indeed be applied to the same specimens on the same optical microscope stage. In many ways, for biological specimens, it complements the other major electronic light microscope techniques of video-enhanced contrast microscopy and low-light-level digital fluorescence imaging using integrating CCD cameras or intensifying video cameras. Its chief advantages are its ability to generate multi-dimensional images by non-invasive optical sectioning with a virtual absence of out-of-focus blur, its capacity for simultaneous multiparameter imaging of multiply-labelled specimens without image registration problems, and its potential for super-resolution.

These advantages have been widely appreciated, and over the last decade have led both to a veritable explosion of demand for confocal microscopes and to a wealth of new research findings as a result of their use. The many different forms of confocal microscopy that have already been developed stand as a tribute to the creativity and inventiveness of the human spirit. However, this branch of optics has only recently emerged from its infancy, and we should expect many more exciting developments within the next decade.

Appendix A

Further reading

Because the intended purpose of this Handbook is to provide a general introduction to confocal microscopy, specific bibliographic citations have been omitted from the text. However, many of the definitive papers describing advances in confocal image acquisition, processing, analysis and display described in this Handbook are to be found in recent issues of the *Journal of Microscopy*.

The following books and reviews are highly recommended for further reading.

Ash, E.A. (ed.) (1980) *Scanned Image Microscopy*. Academic Press, London.

Born, M. and Wolf, E. (1989) *Principles of Optics*, 6th (corrected) Edn. Pergamon Press, Oxford.

Boyde, A., Grey, C. and Jones S. (1997) Real-time confocal microscopy and cell biology. In: *Cell Biology: a Laboratory Handbook*, 2nd Edn (J. Celis, ed.), Vol. 3. Academic Press, San Diego, CA (in press).

Bright, G.R. (1993) Multiparameter imaging of cellular function. In: *Fluorescent and Luminescent Probes for Biological Activity* (W.T. Mason, ed.), pp. 204–215. Academic Press, London.

Castleman, K.R. (1996) *Digital Image Processing*, 2nd Edn. Prentice-Hall Inc, Englewood Cliffs, NJ.

Cheng, P.C. (ed.) (1994) *Computer Assisted Multidimensional Microscopies*. Springer-Verlag, New York.

Haugland, P.R. (ed.) (1996) *Handbook of Fluorescent Probes and Research Chemicals*. 6th Edn. Molecular Probes, Eugene, OR.

Inoué, S. (1986) *Video Microscopy*. Plenum Press, New York. (A second edition is in preparation.)

Kreite, A. (ed.) (1992) *Visualization in Biomedical Microscopies*. VCH, Weinheim.

Ladic, L. (1977) The use of internet graphics software for the processing and display of digital microscopy data. In: *Cell Biology: a Laboratory Handbook*, 2nd Edn (J. Celis, ed.), Vol. 3, Academic Press, San Diego, CA (in press).

Mason, W.T. (ed.) (1993) *Fluorescent and Luminescent Probes for Biological Activity. A Practical Guide to Technology for Quantitative Real-time Analysis.* Academic Press, London.

Matsumoto, B. (ed.) (1993) *Cell Biological Applications of Confocal Microscopy*. Methods in Cell Biology, Vol. 38, Academic Press, San Diego, CA.

Pawley, J.B. (ed.) (1995) *Handbook of Biological Confocal Microscopy*, 2nd Edn. Plenum Press, New York.

Pawley, J.B. and Centonze, V. (1997) Practical laser-scanning confocal microscopy: obtaining optimal performance from your instrument. In: *Cell Biology: a Laboratory Handbook*, 2nd Edn (J. Celis, ed.), Vol. 3. Academic Press, San Diego, CA (in press).

Ploem, J.S. and Tanke, H.J. (1987) *Introduction to Fluorescence Microscopy*. Royal Microscopical Society Microscopy Handbook, No. 10. Oxford University Press, Oxford.

Shaw, P.J. (1997) Computational deblurring of fluorescence microscopy images. In: *Cell Biology: a Laboratory Handbook*, 2nd Edn (J. Celis, ed.), Vol. 3. Academic Press, San Diego, CA (in press).

Sheppard, C.J.R. (1987) Scanning optical microscopy. In: *Advances in Optical and Electron Microscopy* (R. Barer and V.E. Cosslett, eds), Vol. 10, pp. 1–98. Academic Press, London.

Sheppard, C.J.R. (1987) Scanning optical microscopy and its applications in materials science. *Materials Forum* **10**: 94–103.

Sheppard, C.J.R. (1989) Scanning optical microscopy of semiconductor materials and devices. *Scanning Microsc.* **3**: 15–24.

Sheppard, C.J.R. (1990) Fifteen years of scanning microscopy in Oxford. *Proc. R. Microscop. Soc.* **25**: 319–321.

Shotton, D.M. (1989) Confocal scanning optical microscopy and its applications for biological specimens. *J. Cell Sci.* **94**: 175–206.

Shotton, D.M. (ed.) (1993) *Electronic Light Microscopy: the Principles and Practice of Intensified Fluorescence, Video Enhanced Contrast and Confocal Scanning Optical Microscopy*. Techniques in Modern Biomedical Microscopy, Vol. 1. Wiley-Liss, New York.

Shotton, D.M. (1995) Electronic light microscopy – present capabilities and future prospects. *Histochem. Cell Biol.* **104**: 97–137.

Shotton, D.M. (1997) An introduction to electronic light microscopy. In: *Cell Biology: a Laboratory Handbook*, 2nd Edn (J. Celis, ed.), Vol. 3. Academic Press, San Diego, CA (in press).

Shotton, D.M. (1997) Image resolution and digital image processing in electronic light microscopy. In: *Cell Biology: a Laboratory Handbook*, 2nd Edn (J. Celis, ed.), Vol. 3. Academic Press, San Diego, CA (in press).

Whitaker, M. (1997) Fluorescence imaging of living cells. In: *Cell Biology: a Laboratory Handbook*, 2nd Edn (J. Celis, ed.), Vol. 3. Academic Press, San Diego, CA (in press).

Wilson, T. (ed.) (1990) *Confocal Microscopy*. Academic Press, London.

Wilson, T. and Sheppard, C.J.R. (1984) *Theory and Practice of Scanning Optical Microscopy*. Academic Press, London.

Wilson, T. and Sheppar,d C.J.R. (1991) Scanning optical microscopy of microelectronic devices. In: *Analysis of Microelectronic Materials and Devices* (M. Grassbauer and H.W. Werner, eds), pp. 791–810. Wiley, New York.

Wu, J.L. (ed.) (1997) *Focus on Modern Microscopies*. World Scientific, Singapore (in press).

Appendix B

Suppliers

Confocal microscope manufacturers

Bio-Rad Microscience Ltd, Maylands Avenue, Hemel Hempstead, Herts HP2 7TD, UK. Tel: +44 1442 232552; fax: +44 1442 234434.
LASERTEC Corporation, Tsunashimahigashi 4-10-4, Kohoko-Ku, Yokohama, Japan. Tel: +81 45 544 4111; fax: +81 45 543 7764.
Leica UK Ltd, Davy Avenue, Knowlhill, Milton Keynes MK5 8LB, UK. Tel: +44 1908 666663; fax +44 1908 609992.
Meridian Instruments Inc., 2310 Science Parkway, Okemos, MI 48864, USA. Tel: +1 517 349 7200; fax: +1 517 349 5967.
Molecular Dynamics Ltd, 5 Beech House, Asheridge Road, Chesham, Bucks HP5 2PX, UK. Tel: +44 1494 793377; fax: +44 1494 793222.
Nikon Europe B.V., P.O. Box 222, 1170 AE Badhoevedorp, The Netherlands. Tel: +31 20 449 6222; fax +31 20 449 6299.
Noran Instruments UK, GMS House, 26 Rockingham Drive, Linford Wood East, Milton Keynes MK14 6LY, UK. Tel: +44 1908 696290; fax: +44 1908 696292.
Olympus Optical Co. Ltd., 3-1 Nishi Shinjuku 1-Chrome, Shinjuku-Ku, Tokyo 163-09, Japan. Tel: +81 333 772 333; fax: +81 333 756 550.
Optiscan Pty Ltd, PO Box 1066, Mt Waverley, MDC Victoria 3149, Australia. Tel: +61 3 9562 7741; fax: +61 3 9562 7742.
Carl Zeiss, Jena GmbH, Tatzenpromenade 1a, 07745 Jena, Germany. Tel: +49 3641 64 0; fax: +49 3641 64 2856.

Fluorescent filter manufacturers

Chroma Technology Corp., 72 Cotton Mill Hill, Unit A-9, Brattleboro, VT 05301, USA. Tel: +1 802 257 1800; fax: +1 802 257 9400.
Omega Optical Inc., P.O. Box 573, 3 Grove Street, Brattleboro, VT 05302-0573, USA. Tel: +1 802 254 2690; fax: +1 802 254 3937.

3D image processing software suppliers

Bitplane AG, Technoparkstrasse 1, Zürich, Switzerland. Tel: +41 440 2965; fax: +41 1 445 1541.

Data Cell Ltd, Hattori House, Vanwall Business Park, Maidenhead SL6 4UB, UK. Tel: +44 1628 415415; fax: +44 1628 415400.

Foster Findlay Associates Ltd, Newcastle Technopole, Kings Manor, Newcastle upon Tyne NE1 6PA, UK. Tel: +44 191 201 2180; fax: +44 191 201 2190.

Guy Cox Software, P.O. Box 366, Rozelle, NSW 2039, Australia. Tel/fax: +61 2 9818 1896.

Improvision Ltd, Warwick University Science Park, Millburn Hill Road, Coventry CV4 7HS, UK. Tel: +44 203 692229; fax: +44 203 690091.

Kinetic Imaging Ltd, South Harrington Building, Sefton Street, Liverpool L3 4BQ, UK. Tel: +44 151 709 8881; fax: +44 151 709 8633.

Noesis S.A., Immeuble Aviane, Domaine Technologique Saclay, 4 rue René Razel, 91892 Orsay Cedex, France. Tel: +33 1 69 35 30 00; fax: +33 1 69 35 30 01.

Synoptics Ltd, 271, Cambridge Science Park, Milton Road, Cambridge CB4 4WE, UK. Tel: +44 1223 423223; fax: +44 1223 420020.

Vital Images, 3100 West Lake Street, Suite 100, Minneapolis, MN 55416-4510, USA. Tel: +1 612 915 8000; fax: 612 915 8010.

Suppliers of reagents and fluorescent probes

Merck Ltd, Merck House, Poole, Dorset BH15 1TD, UK. Tel: +44 1202 669700; fax: +44 1202 665599.

Calbiochem-Novabiochem (UK) Ltd, Boulevard Industrial Park, Padge Road, Beeston, Nottingham NG9 2JR, UK. Tel: +44 115 943 0840; fax: +44 115 943 0951.

Molecular Probes Inc., 4849 Pitchford Avenue, Eugene, OR 97402-9165, USA. Tel: +1 541 465 8300; fax: +1 541 344 6504.

Sigma-Aldridge Co. Ltd, Fancy Road, Poole, Dorset BH12 4QH, UK. Tel: +44 1202 733114; fax: +44 1202 715460.

Vector Laboratories Inc., 30 Ingold Road, Burlingame, CA 94010, USA. Tel: +1 415 697 3600; fax +1 415 697 0339.

Appendix C

World Wide Web sites relevant to confocal microscopy

A wealth of helpful information relevant to confocal microscopy is available over the Internet. The following World Wide Web sites supplement and extend this Handbook, and will permit the reader to learn from and communicate with the wider confocal community.

List of microscopy sites at Yahoo:

http://www.yahoo.co.uk/Science/Engineering/Optical_Engineering/Microscopy/

WWW Virtual Library, Microscopy:

http://www.ou.edu/research/electron/www-vl/

Microscopy Online magazine: http://www.microscopy-online.com

General information

Microscopes and Microscopy: http://www.lars.bbsrc.ac.uk/micro/
Microscopy and Microanalysis: http://www.amc.anl.gov/

Microscopy societies

Microscopy Society of America: http://www.msa.microscopy.com/
Royal Microscopical Society: http://www.rms.org.uk/
Others: http://www.amc.anl.gov/Docs/NonANL/SocietySites.html

Confocal news groups and discussion lists

Information on how to join confocal and 3D imaging newsgroups, including that for NIH Image, is to be found at:
http://www.ou.edu/research/electron/mirror/newsmail.html

Lance Ladic's confocal and 3D imaging web pages (Ladic, 1997; see Further Reading)

Confocal microscopy resources:
http://www.cs.ubc.ca/spider/ladic/confocal.html
Links to other microscopy sites:
http://www.cs.ubc.ca/spider/ladic/links.html
Sites with confocal images and animations:
http://www.cs.ubc.ca/spider/ladic/conflink.html
Information about image formats:
http://www.cs.ubc.ca/spider/ladic/fileform.html
Freeware/shareware catalogue of software for processing, display and animation of 3D digital microscopy data:
http://www.cs.ubc.ca/spider/ladic/software.html
Making animations for the web:
http://www.cs.ubc.ca/spider/ladic/animate.html

NIH Image (excellent public domain 3D image processing software for Macintosh)

Homepage: http://rsb.info.nih.gov/nih-image/
On-line manual: http://rsb.info.nih.gov/nih-image/manual
Program files to download:
http://rsb.info.nih.gov/nih-image/download.html
Guide for post-acquisition processing of confocal microscopy images:
http://rsb.info.nih.gov/nih-image/more-docs/confocals.html

Index

Aberrations
   chromatic, 28–30
   spherical, 28–29
   with on-axis imaging using
      scanned-stage SOM, 6
Acousto-optic deflector, 8–9
Actin microfilaments, 21–22,
   51–52
α-actinin, 56–57
Airy disc diameter, relative to
   confocal aperture size, 37–38
Airy disc image, 39–40
Anaglyph stereoscopic image
   display, 49, 54–55
Animation of 3D confocal images,
   50–51, 54–55
Antifade agents, 63
Aqueous specimens, aberration-
   free imaging, 28–29
Array detectors, 83
Autofluorescence, 42–43, 48,
   52–53
Avalanche photodiode, 25–26
Average projection, 21–22, 49
Axial imaging, 35
Axial resolution, 2, 6, 33–39, 62
   and confocal aperture diameter,
      36–38
   and integrated intensity, 34
   and numerical aperture, 36
   and point spread function, 33
   and spatial frequency cutoff, 34
   and spherical aberration, 38
   in 4 pi and theta confocal
      microscopy, 85–87

Axial scanning, 18–20
   for surface height
      determination, 77

Beam splitter, for reflection
   imaging, 5, 13
Biopsy specimens, 66
Bleedthrough between
   fluorescence channels, 67

Calcium ion wave, in fertilized
   starfish egg, 58–59
Cardanic scan mirror, 12–13
Cardiac muscle cell, 56–57, 64
Charge coupled device (CCD)
   camera, 3, 8, 24–26, 44
   and scan conversion, 25
Charge coupled device (CCD)
   linear, 75
Chromatic aberration, 28–30
Cilia, 53–55
Coherence of imaging, 5, 12,
   15–18, 75–76
Colloidal gold particles imaged by
   reflection, 18, 71–72
Computational image restoration,
   48, 83
Computer, host for confocal
   imaging system, 46
Condenser, 15–16
Confocal aperture and out-of-focus
   blur removal, 1–2, 4–5
Confocal aperture
   circular 13, 62
   alignment of, 28, 31

and axial resolution, 36–38
and lateral resolution, 40
size for fluorescence imaging, 43
slit, 3, 10, 27, 61–62
and axial resolution, 36–38
and object detectability, 43–44
Confocal fluorescence imaging –
see Fluorescence imaging
Confocal imaging aperture –
see Confocal aperture
Confocal imaging modes –
see Imaging modes
Confocal microscope
performance, 33–44
setup and alignment, 30–31
Confocal principle, 3, 4, 5–6
Confocal reflection imaging –
see Reflection imaging
Cross-talk when imaging at two fluorescence wavelengths, 67

Delayed fluorescence, 69
Dendritic cell, 51–52
Depth discrimination, 76
Depth of field, 19
Depth measurements, rescaling due to refractive index effects, 38
Detectability of object within fluorescent background, 43
Detection
critical, 17
Köhler, 17
Dichroic mirrors
manufacturers of, 97
single and multiple, 4, 5, 13, 61
Digital image processing, 2, 45–52, 62
Drosophila, 59–59

Edge detection, 75–76
Embryos
confocal observation of, 66
observation by 4 pi confocal microscopy, 87
Epidermal leaf cell, 56–57

Epi-illumination, 4, 5, 7, 17–18, 61
Equipment, manufacturers, 97

Fibroblast, 22, 54–55
Fluorescence emissions, 1, 3, 5
and Stokes' shift, 28, 42, 64
incoherence of, 6, 18
two–photon excitation of, 58–59, 87–90
wavelength of, 28
Fluorescence imaging, 1, 61–69
and lateral resolution, 41–42
applications of, 64–66
immunocytochemical labelling for, 28, 47, 52–59, 62–63
instrumentation for, 61–62
specimen preparation techniques, 62–64
Fluorescence *in situ* hybridization (FISH), 65
Fluorescence ratio imaging, 47, 67–68
Fluorescent dextran
for calcium measurement, 56–57
for confocal negative contrast imaging, 64
for photobleaching detection, 89
for pseudo-ratio imaging, 68
Fluorescent microspheres, 18
Fluorochromes, 5, 64
absorption and emission maxima, 65
suppliers, 98
two-photon excitation instead of UV excitation, 90
Four-dimensional imaging, 46
Full-field illumination, 1–2

Golgi-stained neurons, 48, 54–55
Green fluorescent protein (GFP), 56–57, 66
Grey levels, significant, 26

Height profiling, 18–20

Illuminating aperture, 3, 4, 11

Illumination
  critical, 15
  Köhler, 16
Image compression, 46
Image formation, 15–31
Image memory requirements,
  digital, 5, 45, 51
Image plane scanning, 3, 10, 16
Image processing, 45–51
  software suppliers, 98
Image restoration, and super-
  resolution, 48, 83
Image storage, digital, 46
Imaging modes
  confocal darkfield microscopy,
    23
  confocal differential
    interference contrast, 23
  confocal fluorescence, 1, 41–42,
    47, 52–59, 61–69
  confocal interference
    microscopy, 23
  confocal polarization
    miccroscopy, 23
  confocal Raman imaging, 23, 80
  confocal reflection, 1, 3, 17,
    22–23, 41, 71–73
  interference reflection contrast,
    73
  multiparameter fluorescence,
    45, 52–53, 56–59, 61–62, 67
  negative contrast imaging
    using fluorescent dextran,
    64–65
  non-confocal, 22–23
  optical-beam-induced current
    (OBIC), 24, 80–81
  4 pi confocal microscopy, 85–87
  photoluminescence, 80
  theta confocal microscopy,
    85–87
Immunoconjugates, suppliers, 98
Immunocytochemical labelling,
  52–59, 62–63
In situ hybridization, 66
Industrial applications, 75–81

Information content of image,
  25–26
Infrared illumination
  and detector sensitivity, 24
  and two-photon excitation of
    fluorescence, 89
Instrumentation for confocal
  imaging, 61–62
Integrated circuits, 58–59, 76, 78
Interference filters
  manufacturers, 97
  single and multiple, 5, 61, 67
Interference reflection contrast
  imaging, 73
Islet of Langerhans, 58–59

Lasers for confocal microscopy,
  10–12
  coherence of light emission, 5
  diode, for two-photon excitation
    of fluorescence, 90
  dye, for two-photon excitation
    of fluorescence, 90
  frequency-doubled Nd:YAG,
    58–59
  helium–neon, for investigation
    of silicon devices, 80
  multi-line argon–krypton
    mixed gas, 10–11, 67
  neodymium–yttrium
    lanthanum fluoride, 58–59
  pulsed, for two-photon
    excitation of fluorescence, 90
  titanium–sapphire, for two-
    photon excitation, 90
  wavelengths of, 10–11
Lateral resolution, 2, 6, 39–42
  and confocal aperture size, 37
  and confocal fluorescence
    imaging, 41–42
  and confocal reflection imaging,
    41
Light scattering, 42
Line illumination, 8, 10
  and object detectability, 43–44
Linear CCD array, 8, 9, 75

Madin–Darby canine kidney
  epithelial cells, 64
Maximum projection, 21–22, 49,
  54–55
Microtubules, 51–55, 64
Multiparameter imaging, 45,
  52–53, 56–59, 61–62, 67
Multiple-beam scanning, 3, 10
Multiple wavelength imaging,
  52–53, 56–59, 61–62, 67

Nanovid microscopy, 72
Near-field scanning optical
  microscopy (NSOM), 84
Nicotiana, 56–57
Nipkow disc, 3, 10, 62
Noise, 5, 24–27
  dark current, 24, 27
  readout, 24–26
  sensor, 24
Normalized coordinates, 35, 36, 39
Numerical aperture, 27
  and axical resolution, 36
Nyquist sampling frequency, and
  spatial resolution, 46

Object-plane scanning, 3, 10,
  16–17
Objective lenses
  and image resolution, 6, 15,
    27–29
  dry, and axial resolution, 37
  flat-field, 27
  low magnification, 28, 66
  oil-immersion, 27–29
    and axial resolution, 37
    and spherical aberration in
      aqueous specimens, 27–29,
      38–38
  water-immersion, 28–29
    and axial resolution, 37
    and spherical aberration in
      aqueous specimens, 27–29
    use in 4 pi confocal
      microscopy, 86–87
Optical-beam-induced current
  (OBIC), 24, 80–81

Optical fibre, 13
Optical sectioning, 1, 2, 21–22,
  33–40, 48
  by two-photon excitation of
    fluorescence, 88–89
Optical transer function (OTF), 34
Orthogonal projections, 50–51
Out-of-focus blur removal, 1, 4,
  5–6

Paramecium, 48, 54–55
Peak detection algorithm, 77
Perspective projections, 50
Phosphorescence, 69
Photobleaching, 5, 63, 69
  in two-photon excitation of
    fluorescence, 88–89
Photocathode, 24–26
Photodetectors, 24–27
  dynamic range, linearity and
    sensitivity, 24, 44
  wavelength specificity, 24
Photomultiplier, 3, 4, 5, 8, 9, 61
  properties for confocal imaging,
    24–27, 44
Photon counting, 27
Photon-limited imaging of
  fluorescence, 9
Phototoxicity of ultraviolet and
  blue light, 42, 64
  and two-photon excitation of
    fluorescence, 89
4 pi confocal microscopy, 85–87
Piezoelectric z focus drive, 20
Pixel fluorograms, 47, 51–52
Planar specimens, 36
Point objects, 36
Point spread function (PSF), 18,
  40, 48
  and super-resolution, 83–84
  in 4 pi and theta confocal
    microscopy, 85–87
Pollen, 48, 52–53
Potato virus X, 56–57
Prompt fluorescence, 69
Pseudocolour image display, 45,
  47, 51–52, 58–59, 68

Pseudo-ratio imaging, 68

Quantum efficiency, 24

Raster scanning, 3–4
Rayleigh criterion of resolution,
  definition, 40
Reflection imaging, 1, 3, 17,
  22–23
  and lateral resolution, 41
  of colloidal gold particles,
  71–72
  of gold-coated SEM specimens,
  73
  of surface replicas, 72–73
Rhodamine–phalloidin, 21–22,
  51–52

Scan mirrors, 7, 8, 12–13
Scanning modes for SOM
  scanned-beam, 6–7
  scanned-lens, 7
  scanned-stage, 6
Scanning optical microscopy
  (SOM), 3–4
Seedfill algorithm, 48, 54–55
Signal photons, photobleaching
  and saturation, 43–44
Signal-to-noise ratio (SNR), 2, 9,
  25
Significant grey levels, 26
Silicon devices
  non-invasive investigation of,
  79–81
  thin film profiling of, 78
Single-beam scanning, 2–3,
  10–13, 16, 61
Single-photon counting, time-
  correlated, 69
Slit scanning, 3, 10, 27, 61–62
  and object detectability, 44
Slow scan confocal imaging, 6–7
Software for confocal imaging, 62
  suppliers, 98
Space-invariant imaging, 6, 27
Spatial frequency cutoff, 34
Spatial resolution, 9
  and Nyquist sampling
  frequency, 46
  see also – Axial resolution,
  Lateral resolution
Specimen preparation techniques,
  62–64
Spherical aberration
  and axial resolution, 38
  and refractive rescaling of
  depth measurements, 38
  with oil-immersion objective
  lenses, 27–29, 38–39
Stage scanning, 23
Starfish egg fertilization, calcium
  changes, 58–59
Stereology, 48
Stereoscopic image display, 46,
  48–49, 54–55, 72
Stimulated fluorescence emission,
  84
Stokes' shift, 28, 42, 64
Sub-resolution test specimens, 18
Super-resolution, 39–40, 83–84
Surface profiling, 76–78
Surface reflectivity, 58–59, 78
Surface rendering, 50

T lymphocyte, 51–52
Tandem scanning confocal
  microscope, 3, 10
Telecentric lens relay system, 12
Temporal resolution, 9
Theta confocal microscopy, 85–87
Thin film profiling, 78
Three-dimensional (3D) image
  processing, 45–51
  and image memory
  requirements, 46, 51
Threshold projection, 49
Time-lapse confocal imaging, 51
Time-resolved confocal
  fluorescence imaging, 69
Transistor, 81
Transmission detector, 22, 68
Transmission imaging
  confocal, 22–23
  of colloidal gold, 72

non-confocal, 17, 22
   and confocal fluorescence imaging, 68–69
Transverse resolution – *see* Lateral resolution
True confocal imaging, 11, 40
Two-photon excitation of fluorescence, 58–59, 87–90

Ultraviolet illumination, 24, 61, 90
   and phototoxicity, 42, 64

Vertical sectioning, 19–21
Vibration isolation, 30
Video cameras, 3, 8, 10
Video microscopy, 16, 25
Video-rate confocal imaging, 7–10, 27, 61, 62
   and physiological measurements, 66
Volume rendering, 50
Voxels, 50
World Wide Web sites relevant to confocal microscopy, 99–100